Cotton: Origin, History, Technology, and Production

Cotton: Origin, History, Technology, and Production

Prashant Kumar Sirohi

RANDOM PUBLICATIONS
NEW DELHI (INDIA)

Cotton: Origin, History, Technology, and Production

ISBN 978-93-5111-776-6

Published in 2016 in India by

RANDOM PUBLICATIONS

4376-A/4B, Gali Murari Lal, Ansari Road
New Delhi-110 002
Phone : +9111-43580356, 011-23289044, 011-43142548
e-mail: sales@randompublications.com,
info@randompublications.com, randomexports@gmail.com

Type Setting by : Friends Media, Delhi-110089
Digitally Printed at : Replika Press Pvt. Ltd.

Preface

Cotton is a very important crop in India; farmers there face the challenge of losses due to various insect pests. The first genetically modified crop in India, Bt cotton, has been introduced to address bollworm infestation. The process of introduction of Bt cotton took six years of experimentation, during which time agronomic, environmental, and biosafety data was generated and reported.

The trials conducted prior to commercialization clearly established the superior performance of Bt cotton, as demonstrated by increased yields and reduction in application of pesticides. Transgenic technology is suitable for the Indian farmer despite small farm holdings. The area under Bt cotton is projected to increase rapidly in the coming years.

Cotton is more than just a fibre for textiles. It is also an important source of raw materials used in animal feed and for various processed food ingredients. Many countries are now growing genetically modified cotton. In China, GM cotton could drastically reduce pesticide use.

Cotton fibres used in textiles around the world come from the seed hairs of a plant known as Gossypium hirsutum. Cotton, which is cultivated on five continents, develops in closed, green capsules known as bolls that burst open when ripe, revealing the white, fluffy fibres.

India's cotton-processing sector gradually declined during British expansion in India and the establishment of colonial rule during the late 18th and early 19th centuries. This was largely due to aggressive colonialist mercantile policies of the British East India Company, which made cotton processing and manufacturing workshops in India uncompetitive. Indian markets were increasingly forced to supply only raw cotton and were forced, by British-imposed law, to purchase manufactured textiles from Britain.

With a comprehensive subject index that makes reference hunting easier, appended to the book, it is a work of lasting importance of scholars, researchers, agricultural scientists, cotton growers as well as all those linked with cotton industry and cotton marketing.

– ***Author***

Contents

1

Introduction

Cotton is a soft, fluffy staple fiber that grows in a boll, or protective case, around the seeds of cotton plants of the genus *Gossypium* in the family of *Malvaceae*. The fiber is almost pure cellulose. Under natural conditions, the cotton bolls will tend to increase the dispersal of the seeds.

The plant is a shrub native to tropical and subtropical regions around the world, including the Americas, Africa, and India. The greatest diversity of wild cotton species is found in Mexico, followed by Australia and Africa. Cotton was independently domesticated in the Old and New Worlds.

The English name derives from the Arabic *(al) qum□n*, which began to be used circa 1400 AD.

The fiber is most often spun into yarn or thread and used to make a soft, breathable textile. The use of cotton for fabric is known to date to prehistoric times; fragments of cotton fabric dated from 5000 BC have been excavated in Mexico and the Indus Valley Civilization in Ancient India (modern-day Pakistan and some parts of India).

Although cultivated since antiquity, it was the invention of the cotton gin that lowered the cost of production that led to its widespread use, and it is the most widely used natural fiber cloth in clothing today.

Current estimates for world production are about 25 million tonnes or 110 million bales annually, accounting for 2.5% of the world's arable land. China is the world's largest producer of cotton, but most of this is used domestically. The United States has been the largest exporter for many years.

In the United States, cotton is usually measured in bales, which measure approximately 0.48 cubic metres (17 cubic feet) and weigh 226.8 kilograms (500 pounds).

NATURE OF COTTON

Botanically, three principal groups of cotton are of commercial importance. The first, the species Gossypium hirsutum, is native to Mexico and Central America and has been developed for extensive use in the United States, accounting for more than ninety-five percent of U.S. production. This group is known in the United States as "American Upland" cotton and has fibers that range in length from about 7/8 to 15/16 inches. The second botanical group, the species G. barbadense, which makes up the balance of U.S. production, is of early South American origin. With fibers ranging in length from 1¼ inches to 19/16 inches, it is known in the United States as "American Pima" cotton, also commonly referred to as "Extra-Long Staple" cotton. A third group, G. herbaceum and G. arboreum, consists of cottons with shorter fiber lengths, ½ to 1 inch, that are native to India and Eastern Asia. No cottons from this group are grown in the United States.

A single pound of cotton may contain 100 million or more individual fibers. Each fiber is an outgrowth of a single cell that develops in the surface layer of the cotton seed. During early stages of its growth, the fiber elongates to its full length as a thin-walled tube. As it matures, the fiber wall is thickened by deposits of cellulose inside the tube, leaving a hollow area in the center. When the growth period ends and the living material dies, the fiber collapses and twists about its own axis.

CLASSIFICATION

The term "cotton classification" in this publication refers to the application of official standards and standardized procedures developed by USDA for measuring those physical attributes of raw cotton that affect the quality of the finished product and/or manufacturing efficiency. USDA's classing methodology is based on both grade and instrument standards used hand-in-hand with state-of-the-art methods and equipment to provide the cotton industry with the best possible information on cotton quality for marketing and processing. USDA classification currently consists of determinations of fiber length, length uniformity, fiber strength, micronaire, color, trash, leaf, and extraneous matter.

Fig. *USDA's classing methodology is based on both grade and instrument standards used hand-in-hand with state-of-the-art methods and equipment.*

The system is rapidly moving from reliance on the human senses to the use of high-volume, precision instruments that perform quality measurements in a matter of seconds. Only the classifications for extraneous matter and special conditions are still performed manually. Research and development continue for the technology and instrumentation to rapidly measure extraneous matter, as well as other important fiber characteristics, such as maturity, stickiness, short-fiber content, and neps. USDA will complete the transition to an allinstrument classification as quickly as the technology can be developed and instruments are sufficiently refined to assure representative and reliable quality measurements.

STRUCTURE

USDA provides cotton classification services under the direction of the Agricultural Marketing Service (AMS) Cotton and Tobacco Program. The Program has eight main areas of operation: the Grading Division, the Standardization and Engineering Division, the Quality Assurance Division, the Market News Division, the Information Technology Division, the Research and Promotion Staff, the Administrative Staff, and the Program Appraisal Staff. Each area of operation plays an integral role in maintaining a reliable, efficient, and effective classification system and delivery of services.

SCOPE

Practically all cotton grown in the United States is classed by USDA at the request of producers. Although classification is not mandatory, growers generally find it essential to marketing their crop and for participation in the USDA price support program. The USDA AMS Cotton and Tobacco Program operates ten cotton classing facilities across the Cotton Belt (their locations are shown on the map on the inside back cover of this booklet). These facilities, which are part of Grading Division, are designed specifically for cotton classification and are staffed exclusively with USDA personnel.

Fig. *The gin stand separates the cotton fibers from the seed.*

USDA also classes all cotton tendered for delivery on futures contracts on the Intercontinental Exchange and provides arbitration classing to the industry. These services are performed by the Quality Assurance Division. Classification services are also provided to individual buyers, manufacturers, breeders, researchers, and others upon request. All users of USDA classification services are charged fees to recover classification costs.

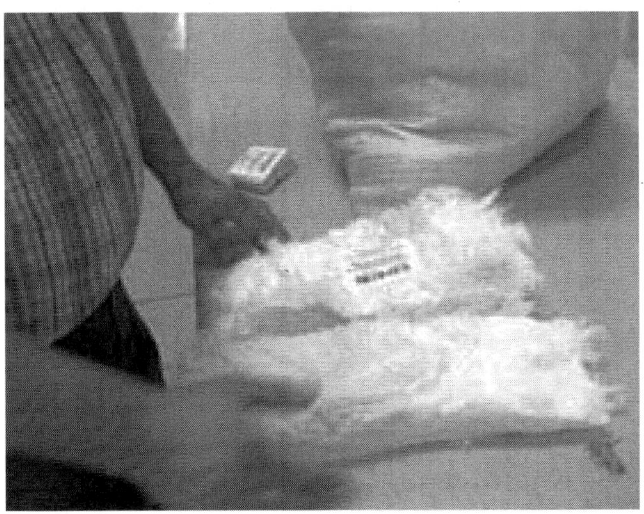

Fig. *A 4-ounce sample is taken from each side of the bale by a licensed sampling agent and forwarded to USDA for classification.*

PROCESS

At the gin, cotton fibers are separated from the seed, cleaned to remove plant residue and other foreign material, and pressed into bales of approximately 500 pounds. A sample of at least 4 ounces (115 grams) is taken from each side of the bale by a licensed sampling agent and identified with a Permanent Bale Identification (PBI) tag. The total 8 ounce (230gram) sample is delivered by the agent or a designated hauler to the USDA classing facility serving the area. Gin and warehouse operators serve as licensed sampling agents and perform this function under USDA supervision.

Upon arrival at the classing facility, the samples are conditioned to bring the moisture content into a specified range before the classing process begins. The samples are then transported to the instrument-testing and manual-classing stations, where classification is performed. Remnants of the samples used during the classification process are baled and sold by USDA, with the proceeds applied to offset classification costs.

Once classification is complete, the fiber measurement results are immediately available to the customer from the classing facility's database. Providing cotton quality results quickly gives producers and buyers access to crucial information at the time of sale. At the peak of the season, USDA classes and provides data on as many as two million bales per week nationwide.

TRACKING

The PBI system allows cotton to be tracked from the field to the classing office. On the field, each cotton module is labeled with an identification number that links it to the producer, field, and seed variety. At the gin, each module identification number is logged into a database, and each bale is labeled with a PBI tag with a twelve-digit number and barcode identifying the classing office, gin, and bale. Samples taken at the gin for classing also are labeled with PBI tags. At the classing office, the PBI tag follows the sample through testing. The results are linked to the bale and stored in the USDA AMS Cotton and Tobacco Program's National Database by PBI number. The classification data in the National Database can be accessed by the owner of the cotton or the owner's authorized agent. Users of this system include grower marketing cooperatives, buyers, and textile manufacturers.

OVERVIEW OF THE COTTON CLASSIFICATION PROCESS

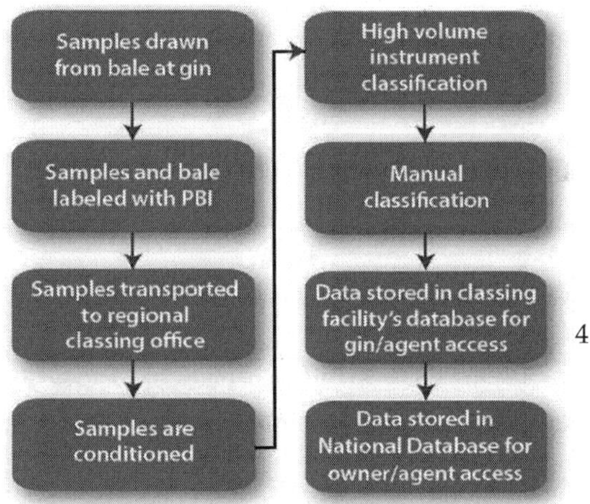

4

TYPES

There are four commercially grown species of cotton, all domesticated in antiquity:

- *Gossypium hirsutum* – upland cotton, native to Central America, Mexico, the Caribbean and southern Florida (90% of world production)
- *Gossypium barbadense* – known as extra-long staple cotton, native to tropical South America (8% of world production)
- *Gossypium arboreum* – tree cotton, native to India and Pakistan (less than 2%)

- *Gossypium herbaceum* – Levant cotton, native to southern Africa and the Arabian Peninsula (less than 2%)

The two New World cotton species account for the vast majority of modern cotton production, but the two Old World species were widely used before the 1900s. While cotton fibers occur naturally in colors of white, brown, pink and green, fears of contaminating the genetics of white cotton have led many cotton-growing locations to ban the growing of colored cotton varieties, which remain a specialty product.

HISTORY

Fig. Cotton plants as imagined and drawn by John Mandeville in the 14th century

Cotton was used in the Old World at least 7,000 years ago (5th millennium BC). Evidence of cotton use has been found at the site ofMehrgarh, where early cotton threads have been preserved in copper beads. Cotton cultivation became more widespread during theIndus Valley Civilization, which covered parts of modern eastern Pakistan and northwestern India. The Indus cotton industry was well-developed and some methods used in cotton spinning and fabrication continued to be used until the industrialization of India. Between 2000 and 1000 BC cotton became widespread across much of India. For example, it has been found at the site of Hallus in Karnatakadating from around 1000 BC. Cotton fabrics discovered in a cave near Tehuacán, Mexico have been dated to around 5800 BC, although it is difficult to know for certain due to fiber decay. Other sources date the domestication of cotton in Mexico to approximately 5000 to 3000 BC.

The Greeks and the Arabs were not familiar with cotton until the Wars of Alexander the Great, as his contemporary Megasthenes toldSeleucus I Nicator of "there being trees on which wool grows" in "Indica". This may be a reference to "tree cotton",Gossypium arboreum, which is a native of the Indian subcontinent.

According to the *Columbia Encyclopedia*: Cotton has been spun, woven, and dyed since prehistoric times. It clothed the people of ancient India, Egypt, and China. Hundreds of years before the Christian era, cotton textiles were woven in India with matchless skill, and their use spread to the Mediterranean countries.

In Iran (Persia), the history of cotton dates back to the Achaemenid era (5th century BC); however, there are few sources about the planting of cotton in pre-Islamic Iran.

The planting of cotton was common in Merv, Ray and Pars of Iran. In Persian poets' poems, especially Ferdowsi's Shahname, there are references to cotton ("panbe" in Persian).Marco Polo (13th century) refers to the major products of Persia, including cotton. John Chardin, a French traveler of the 17th century who visited the Safavid Persia, spoke approvingly of the vast cotton farms of Persia. During the Han dynasty, cotton was grown by non-Chinese peoples in the southern Chinese province of Yunnan.

In Peru, cultivation of the indigenous cotton species *Gossypium barbadense* was the backbone of the development of coastal cultures such as the Norte Chico, Moche, andNazca. Cotton was grown upriver, made into nets, and traded with fishing villages along the coast for large supplies of fish. The Spanish who came to Mexico and Peru in the early 16th century found the people growing cotton and wearing clothing made of it.

During the late medieval period, cotton became known as an imported fiber in northern Europe, without any knowledge of how it was derived, other than that it was a plant. Because Herodotus had written in his *Histories*, Book III, 106, that in India trees grew in the wild producing wool, it was assumed that the plant was a tree, rather than a shrub. This aspect is retained in the name for cotton in several Germanic languages, such as German *Baumwolle*, which translates as "tree wool" (*Baum* means "tree"; *Wolle* means "wool"). Noting its similarities to wool, people in the region could only imagine that cotton must be produced by plant-borne sheep. John Mandeville, writing in 1350, stated as fact the now-preposterous belief: "There grew there [India] a wonderful tree which bore tiny lambs on the endes of its branches. These branches were so pliable that they bent down to allow the lambs to feed when they are hungrie [*sic*]." By the end of the 16th century, cotton was cultivated throughout the warmer regions in Asia and the Americas.

India's cotton-processing sector gradually declined during British expansion in India and the establishment of colonial rule during the late 18th and early 19th centuries. This was largely due to aggressive colonialist mercantile policies of the British East India Company, which made cotton processing and manufacturing workshops in India uncompetitive. Indian markets were increasingly forced to supply only raw cotton and were forced, by British-imposed law, to purchase manufactured textiles from Britain.

Fig. The Vegetable Lamb of Tartary

INDUSTRIAL REVOLUTION IN BRITAIN

The advent of the Industrial Revolution in Britain provided a great boost to cotton manufacture, as textiles emerged as Britain's leading export. In 1738, Lewis Paul and John Wyatt, of Birmingham, England, patented the roller spinning machine, as well as the flyer-and-bobbin system for drawing cotton to a more even thickness using two sets of rollers that traveled at different speeds. Later, the invention of the James Hargreaves' spinning jenny in 1764, Richard Arkwright's spinning frame in 1769 and Samuel Crompton's spinning mule in 1775 enabled British spinners to produce cotton yarn at much higher rates. From the late 18th century on, the British city of Manchesteracquired the nickname *"Cottonopolis"* due to the cotton industry's omnipresence within the city, and Manchester's role as the heart of the global cotton trade.

Production capacity in Britain and the United States was improved by the invention of the cotton gin by the American Eli Whitney in 1793. Before the development of cotton gins, the cotton fibers had to be pulled from the seeds tediously by hand. By the late 1700s a number of crude ginning machines had been developed. However, to produce a bale of cotton required over 600 hours of human labor, making large-scale production uneconomical in the United States, even with the use of humans as slave labor. The gin that Whitney manufactured (the Holmes design) reduced the hours down to just a dozen or so per bale. Although Whitney patented his own design for a cotton gin, he manufactured a prior design from Henry Odgen Holmes, for which Holmes filed a patent in 1796.Improving technology and increasing control of world markets allowed British traders to develop a commercial chain in which raw cotton fibers were (at first) purchased fromcolonial plantations, processed into cotton cloth in the mills of Lancashire, and then exported on British ships to captive colonial markets in West Africa, India, and China (via Shanghai and Hong Kong).

By the 1840s, India was no longer capable of supplying the vast quantities of cotton fibers needed by mechanized British factories, while shipping bulky, low-price cotton from India to Britain was time-consuming and expensive. This, coupled with the emergence of American cotton as a superior type (due to the longer, stronger fibers of the two domesticated native American species, *Gossypium hirsutum* and *Gossypium barbadense*), encouraged British traders to purchase cotton from plantations in the United Statesand plantations in the Caribbean. By the mid-19th century, "King Cotton" had become the backbone of the southern American economy. In the United States, cultivating and harvesting cotton became the leading occupation of slaves.

During the American Civil War, American cotton exports slumped due to a Union blockade on Southern ports, and also because of a strategic decision by the Confederategovernment to cut exports, hoping to force Britain to recognize the Confederacy or enter the war. This prompted the main purchasers of cotton, Britain and France, to turn toEgyptian cotton. British and French traders invested heavily in cotton plantations. The Egyptian government of Viceroy Isma'il took out substantial loans from European bankers and stock exchanges. After the American Civil War ended in 1865, British and French traders abandoned Egyptian cotton and returned to cheap American exports,sending Egypt into a deficit spiral that led to the country declaring bankruptcy in 1876, a key factor behind Egypt's occupation by the British Empire in 1882.

Fig. Prisoners farming cotton under thetrusty system in Parchman Farm,Mississippi, 1911

During this time, cotton cultivation in the British Empire, especially India, greatly increased to replace the lost production of the American South. Through tariffs and other restrictions, the British government discouraged the production of cotton cloth in India; rather, the raw fiber was sent to England for processing. The Indian Mahatma Gandhi described the process:

1. English people buy Indian cotton in the field, picked by Indian labor at seven cents a day, through an optional monopoly.
2. This cotton is shipped on British ships, a three-week journey across the Indian Ocean, down the Red Sea, across the Mediterranean, through Gibraltar, across the Bay of Biscay and the Atlantic Ocean to London. One hundred per cent profit on this freight is regarded as small.
3. The cotton is turned into cloth in Lancashire. You pay shilling wages instead of Indian pennies to your workers. The English worker not only has the advantage of better wages, but the steel companies of England get the profit of building the factories and machines. Wages; profits; all these are spent in England.
4. The finished product is sent back to India at European shipping rates, once again on British ships. The captains, officers, sailors of these ships, whose wages must be paid, are English. The only Indians who profit are a few lascars who do the dirty work on the boats for a few cents a day.
5. The cloth is finally sold back to the kings and landlords of India who got the money to buy this expensive cloth out of the poor peasants of India who worked at seven cents a day.

In the United States, Southern cotton provided capital for the continuing development of the North. The cotton produced by enslaved African Americans not only helped the South, but also enriched Northern merchants. Much of the Southern cotton was trans-shipped through northern ports.

Cotton remained a key crop in the Southern economy after emancipation and the end of the Civil War in 1865. Across the South, sharecropping evolved, in which landless black and white farmers worked land owned by others in return for a share of the profits. Some farmers rented the land and bore the production costs themselves.

Until mechanical cotton pickers were developed, cotton farmers needed additional labor to hand-pick cotton. Picking cotton was a source of income for families across the South. Rural and small town school systems had split vacations so children could work in the fields during "cotton-picking."

It was not until the 1950s that reliable harvesting machinery was introduced (prior to this, cotton-harvesting machinery had been too clumsy to pick cotton without shredding the fibers). During the first half of the 20th century, employment in the cotton industry fell, as machines began to replace laborers and the South's rural labor force dwindled during the World Wars.

Cotton remains a major export of the southern United States, and a majority of the world's annual cotton crop is of the long-staple American variety.

CULTIVATION

Fig. Cotton plowing in Togo, 1928

Fig. Picking cotton in Armenia in the 1930s. No cotton is grown there today.

Fig. Cotton modules in Australia (2007)

Successful cultivation of cotton requires a long frost-free period, plenty of sunshine, and a moderate rainfall, usually from 600 to 1,200 mm (24 to 47 in). Soils usually need to be fairly heavy, although the level of nutrients does not need to be exceptional. In general, these conditions are met within the seasonally dry tropics and subtropics in the Northern and Southern hemispheres, but a large proportion of the cotton grown today is cultivated in areas with less rainfall that obtain the water from irrigation. Production of the crop for a given year usually starts soon after harvesting the preceding autumn. Cotton is naturally a perennial but is grown as an annual to help control pests. Planting time in spring in the Northern hemisphere varies from the beginning of February to the beginning of June. The area of the United States known as the South Plains is the largest contiguous cotton-growing region in the world. While dryland (non-irrigated) cotton is successfully grown in this region, consistent yields are only produced with heavy reliance on irrigation water drawn from the Ogallala Aquifer. Since cotton is somewhat salt and drought tolerant, this makes it an attractive crop for arid and semiarid regions. As water resources get tighter around the world, economies that rely on it face difficulties and conflict, as well as potential environmental problems. For example, improper cropping and irrigation practices have led to desertification in areas ofUzbekistan, where cotton is a major export. In the days of the Soviet Union, the Aral Sea was tapped for agricultural irrigation, largely of cotton, and now salination is widespread. Cotton can also be cultivated to have colors other than the yellowish off-white typical of modern commercial cotton fibers. Naturally colored cotton can come in red, green, and several shades of brown.

GENETIC MODIFICATION

Genetically modified (GM) cotton was developed to reduce the heavy reliance on pesticides. The bacterium *Bacillus thuringiensis* (Bt) naturally produces a chemical harmful only to a small fraction of insects, most notably the larvae of moths and butterflies, beetles, andflies, and harmless to other forms of life. The gene coding for Bt toxin has been inserted into cotton, causing cotton, called Bt cotton, to produce this natural insecticide in its tissues. In many regions, the main pests in commercial cotton are lepidopteran larvae, which are killed by the Bt protein in the transgenic cotton they eat. This eliminates the need to use large amounts of broad-spectrum insecticides to kill lepidopteran pests (some of which have developed pyrethroid resistance). This spares natural insect predators in the farm ecology and further contributes to noninsecticide pest management.

But Bt cotton is ineffective against many cotton pests, however, such as plant bugs, stink bugs, and aphids; depending on circumstances it may still be desirable to use insecticides against these. A 2006 study done by Cornell researchers, the Center for Chinese Agricultural Policy and the Chinese

Academy of Science on Bt cotton farming in China found that after seven years these secondary pests that were normally controlled by pesticide had increased, necessitating the use of pesticides at similar levels to non-Bt cotton and causing less profit for farmers because of the extra expense of GM seeds. However, a 2009 study by the Chinese Academy of Sciences, Stanford University and Rutgers University refuted this. They concluded that the GM cotton effectively controlled bollworm. The secondary pests were mostly miridae (plant bugs) whose increase was related to local temperature and rainfall and only continued to increase in half the villages studied. Moreover, the increase in insecticide use for the control of these secondary insects was far smaller than the reduction in total insecticide use due to Bt cotton adoption. A 2012 Chinese study concluded that Bt cotton halved the use of pesticides and doubled the level of ladybirds, lacewings and spiders. The International Service for the Acquisition of Agri-biotech Applications(ISAAA) said that, worldwide, GM cotton was planted on an area of 25 million hectares in 2011. This was 69% of the worldwide total area planted in cotton.

GM cotton acreage in India grew at a rapid rate, increasing from 50,000 hectares in 2002 to 10.6 million hectares in 2011. The total cotton area in India was 12.1 million hectares in 2011, so GM cotton was grown on 88% of the cotton area. This made India the country with the largest area of GM cotton in the world. A long-term study on the economic impacts of Bt cotton in India, published in the Journal PNAS in 2012, showed that Bt cotton has increased yields, profits, and living standards of smallholder farmers. The U.S. GM cotton crop was 4.0 million hectares in 2011 the second largest area in the world, the Chinese GM cotton crop was third largest by area with 3.9 million hectares and Pakistan had the fourth largest GM cotton crop area of 2.6 million hectares in 2011. The initial introduction of GM cotton proved to be a success in Australia – the yields were equivalent to the non-transgenic varieties and the crop used much less pesticide to produce (85% reduction). The subsequent introduction of a second variety of GM cotton led to increases in GM cotton production until 95% of the Australian cotton crop was GM in 2009 making Australia the country with the fifth largest GM cotton crop in the world. Other GM cotton growing countries in 2011 were Argentina, Myanmar, Burkina Faso, Brazil, Mexico, Colombia, South Africa and Costa Rica.

Cotton has been genetically modified for resistance to glyphosate a broad-spectrum herbicide discovered by Monsanto which also sells some of the Bt cotton seeds to farmers. There are also a number of other cotton seed companies selling GM cotton around the world. About 62% of the GM cotton grown from 1996 to 2011 was insect resistant, 24%stacked product and 14% herbicide resistant. Cotton has gossypol, a toxin that makes it inedible. However, scientists have silenced the gene that produces the toxin, making it a potential food crop.

ORGANIC PRODUCTION

Organic cotton is generally understood as cotton from plants not genetically modified and that is certified to be grown without the use of any synthetic agricultural chemicals, such as fertilizers or pesticides. Its production also promotes and enhances biodiversity and biological cycles. United States cotton plantations are required to enforce the National Organic Program (NOP). This institution determines the allowed practices for pest control, growing, fertilizing, and handling of organic crops. As of 2007, 265,517 bales of organic cotton were produced in 24 countries, and worldwide production was growing at a rate of more than 50% per year.

PESTS AND WEEDS

Main article: List of cotton diseases

Fig. Hoeing a cotton field to remove weeds, Greene County, Georgia, USA, 1941

Fig. Female and nymph Cotton Harlequin Bug

The cotton industry relies heavily on chemicals, such as herbicides, fertilizers and insecticides, although a very small number of farmers are moving toward an organic model of production, and organic cotton products are now available for purchase at limited locations. These are popular for baby clothes and diapers. Under most definitions, organic products do not use genetic engineering. All natural cotton products are known to be both sustainable and hypoallergenic.

Historically, in North America, one of the most economically destructive pests in cotton production has been the boll weevil. Due to the US Department of Agriculture's highly successful Boll Weevil Eradication Program (BWEP), this pest has been eliminated from cotton in most of the United States. This program, along with the introduction of genetically engineered Bt cotton (which contains a bacterial gene that codes for a plant-produced protein that is toxic to a number of pests such as cotton bollworm and pink bollworm), has allowed a reduction in the use of synthetic insecticides.

Other significant global pests of cotton include the pink bollworm, *Pectinophora gossypiella*; the chili thrips, *Scirtothrips dorsalis*; the cotton seed bug, *Oxycarenus hyalinipennis*; the tarnish plant bug, *Lygus lineolaris*; and the fall armyworm, *Spodoptera frugiperda,Xanthomonas citri subsp. malvacearum*.

HARVESTING

Fig. Offloading freshly harvested cotton into a module builder in Texas; previously built modules can be seen in the background

Most cotton in the United States, Europe, and Australia is harvested mechanically, either by a cotton picker, a machine that removes the cotton from the boll without damaging the cotton plant, or by a cotton stripper, which strips the entire boll off the plant. Cotton strippers are used in regions where it is too windy to grow picker varieties of cotton, and usually after application of a chemical defoliant or the natural defoliation that occurs after a freeze. Cotton is a perennial crop in the tropics, and without defoliation or freezing, the plant will continue to grow.

Cotton continues to be picked by hand in developing countries.

Fig. Cotton being picked by hand inIndia, 2005.

COMPETITION FROM SYNTHETIC FIBERS

The era of manufactured fibers began with the development of rayon in France in the 1890s. Rayon is derived from a natural cellulose and cannot be considered synthetic, but requires extensive processing in a manufacturing process, and led the less expensive replacement of more naturally derived materials. A succession of new synthetic fibers were introduced by the chemicals industry in the following decades. Acetate in fiber form was developed in 1924. Nylon, the first fiber synthesized entirely from petrochemicals, was introduced as a sewing thread by DuPont in 1936, followed by DuPont's acrylic in 1944. Some garments were created from fabrics based on these fibers, such as women's hosiery from nylon, but it was not until the introduction of polyester into the fiber marketplace in the early 1950s that the market for cotton came under threat. The rapid uptake of polyester garments in the 1960s caused economic hardship in cotton-exporting economies, especially in Central American countries, such as Nicaragua, where cotton production had boomed tenfold between 1950 and 1965 with the advent of cheap chemical pesticides. Cotton production recovered in the 1970s, but crashed to pre-1960 levels in the early 1990s.

Beginning as a self-help program in the mid-1960s, the Cotton Research and Promotion Program (CRPP) was organized by U.S. cotton producers in response to cotton's steady decline in market share. At that time, producers voted to set up a per-bale assessment system to fund the program, with built-in safeguards to protect their investments. With the passage of the Cotton Research and Promotion Act of 1966, the program joined forces and began battling synthetic competitors and re-establishing markets for cotton. Today, the success of this program has made cotton the best-selling fiber in the U.S. and one of the best-selling fibers in the world.

Administered by the Cotton Board and conducted by Cotton Incorporated, the CRPP works to greatly increase the demand for and profitability of cotton through various research and promotion activities. It is funded by U.S. cotton producers and importers.

USES

Cotton is used to make a number of textile products. These include terrycloth for highly absorbent bath towels and robes; denim for blue jeans; cambric, popularly used in the manufacture of blue work shirts (from which we get the term "blue-collar"); and corduroy, seersucker, and cotton twill. Socks, underwear, and most T-shirts are made from cotton. Bed sheets often are made from cotton. Cotton also is used to make yarn used in crochet and knitting. Fabric also can be made from recycled or recovered cotton that otherwise would be thrown away during the spinning, weaving, or cutting process. While many fabrics are made completely of cotton, some materials blend cotton with other fibers, including rayon and synthetic fibers such as polyester. It can either be used in knitted or woven fabrics, as it can be blended with elastine to make a stretchier thread for knitted fabrics, and apparel such as stretch jeans.

In addition to the textile industry, cotton is used in fishing nets, coffee filters, tents, explosives manufacture, cotton paper, and in bookbinding. The first Chinese paper was made of cotton fiber. Fire hoses were once made of cotton.

The cottonseed which remains after the cotton is ginned is used to produce cottonseed oil, which, after refining, can be consumed by humans like any other vegetable oil. Thecottonseed meal that is left generally is fed to ruminant livestock; the gossypol remaining in the meal is toxic to monogastric animals. Cottonseed hulls can be added to dairy cattle rations for roughage. During the American slavery period, cotton root bark was used in folk remedies as an abortifacient, that is, to induce a miscarriage. Gossypol was one of the many substances found in all parts of the cotton plant and it was described by the scientists as 'poisonous pigment'. It also appears to inhibit the development of sperm or even restrict the mobility of the sperm. Also, it is thought to interfere with the menstrual cycle by restricting the release of certain hormones.

Cotton linters are fine, silky fibers which adhere to the seeds of the cotton plant after ginning. These curly fibers typically are less than D_8 inch (3.2 mm) long. The term also may apply to the longer textile fiber staple lint as well as the shorter fuzzy fibers from some upland species. Linters are traditionally used in the manufacture of paper and as a raw material in the manufacture of cellulose. In the UK, linters are referred to as "cotton wool". This can also be a refined product (*absorbent cotton* in U.S. usage) which hasmedical, cosmetic and many other practical uses. The first medical use of cotton wool was by Sampson Gamgee at the Queen's Hospital (later the General Hospital) inBirmingham, England. Shiny cotton is a processed version of the fiber that can be made into cloth resembling satin for shirts and suits. However, it is hydrophobic (does not absorb water easily), which makes it unfit for use in bath and dish towels (although examples of these made from shiny cotton are seen). The name Egyptian cotton is broadly associated with quality products, however only a small percentage of Egyptian cotton production is actually of superior quality. Most products bearing the name are not made with the finest cottons from Egypt.

Pima cotton is often compared to Egyptian cotton, as both are used in high quality bed sheets and other cotton products. It is considered the next best quality after high quality Egyptian cotton by some authorities. Pima cotton is grown in the American southwest. Not all products bearing the Pima name are made with the finest cotton. The Pima name is now used by cotton-producing nations such as Peru, Australia and Israel.

Cotton lisle is a finely-spun, tightly twisted type of cotton that is noted for being stong and durable. Lisle is composed of two strands that have each been twisted an extra twist per inch than ordinary yarns and combined to create a single thread. The yarn is spun so that it is compact and solid. This cotton is used mainly for underwear, stockings, and gloves. Colors applied to this yarn are noted for being more brilliant than colors applied to softer yarn. This type of thread was first made in the city of Lisle, France (now Lille), hence its name.

INTERNATIONAL TRADE

The largest producers of cotton, currently (2009), are China and India, with annual production of about 34 million bales and 27 million bales, respectively; most of this production is consumed by their respective textile industries. The largest exporters of raw cotton are the United States, with sales of $4.9 billion, and Africa, with sales of $2.1 billion. The total international trade is estimated to be $12 billion. Africa's share of the cotton trade has doubled since 1980. Neither area has a significant domestic textile industry, textile manufacturing having moved to developing nations in Eastern and South Asia such as India and China. In Africa, cotton is grown by numerous small holders. Dunavant Enterprises, based in Memphis, Tennessee, is the leading cotton broker in Africa, with hundreds of purchasing agents. It operates cotton gins in Uganda, Mozambique, and Zambia. In Zambia, it often offers loans for seed and expenses to the 180,000 small farmers who grow cotton for it, as well as advice on farming methods. Cargill also purchases cotton in Africa for export.

The 25,000 cotton growers in the United States of America are heavily subsidized at the rate of $2 billion per year although China now provides the highest overall level of cotton sector support. The future of these subsidies is uncertain and has led to anticipatory expansion of cotton brokers' operations in Africa. Dunavant expanded in Africa by buying out local operations. This is only possible in former British colonies and Mozambique; former French colonies continue to maintain tight monopolies, inherited from their former colonialist masters, on cotton purchases at low fixed prices.

LEADING PRODUCER COUNTRIES

Top 10 Cotton Producing Countries (in metric tonnes)				
Rank	**Country**	**2010**	**2011**	**2012**
1	China	5,970,000	6,588,950	6,840,000
2	India	5,683,000	5,984,000	5,321,000
3	United States	3,941,700	3,412,550	3,598,000
4	Pakistan	1,869,000	2,312,000	2,215,000
5	Brazil	973,449	1,673,337	1,638,103
6	Uzbekistan	1,136,120	983,400	1,052,000
7	Turkey	816,705	954,600	851,000
8	Australia	386,800	843,572	973,497
9	Argentina	230,000	295,000	210,000
10	Turkmenistan	225,000	195,000	198,000
–	*World*	22,714,154	24,941,738	25,955,096

The five leading exporters of cotton in 2011 are (1) the United States, (2) India, (3) Brazil, (4) Australia, and (5) Uzbekistan. The largest nonproducing importers are Korea, Taiwan, Russia, and Japan.

In India, the states of Maharashtra (26.63%), Gujarat (17.96%) and Andhra Pradesh (13.75%) and also Madhya Pradesh are the leading cotton producing states, these states have a predominantly tropical wet and dry climate.

In Pakistan, cotton is grown predominantly in the provinces of Punjab, and Sindh. The leading area of cotton production is the south Punjab, comprising the areas around Rahim Yar Khan, Bahawalpur,Bahawalnagar, Multan, Dera Ghazi Khan, Muzaffargarh, Vehari, and Khanewal. In Sindh Sanghar is the most important cotton producing district. Faisalabad is a leader in textiles within Pakistan. Punjab has a tropical wet and dry climate throughout the year therefore enhancing the growth of cotton.

In the United States, the state of Texas led in total production as of 2004, while the state of Californiahad the highest yield per acre.

FAIR TRADE

Cotton is an enormously important commodity throughout the world. However, many farmers in developing countries receive a low price for their produce, or find it difficult to compete with developed countries.

This has led to an international dispute: On 27 September 2002, Brazil requested consultations with the US regarding prohibited and actionable subsidies provided to US producers, users and/or exporters of upland cotton, as well as legislation, regulations, statutory instruments and amendments thereto providing such subsidies (including export credits), grants, and any other assistance to the US producers, users and exporters of upland cotton.

On 8 September 2004, the Panel Report recommended that the United States "withdraw" export credit guarantees and payments to domestic users and exporters, and "take appropriate steps to remove the adverse effects or withdraw" the mandatory price-contingent subsidy measures.

While Brazil was fighting the US through the WTO's Dispute Settlement Mechanism against a heavily subsidized cotton industry, a group of four least-developed African countries – Benin, Burkina Faso, Chad, and Mali – also known as "Cotton-4" have been the leading protagonist for the reduction of US cotton subsidies through negotiations. The four introduced a "Sectoral Initiative in Favour of Cotton", presented by Burkina Faso's President Blaise Compaoré during the Trade Negotiations Committee on 10 June 2003.

In addition to concerns over subsidies, the cotton industries of some countries are criticized for employing child labor and damaging workers' health by exposure to pesticides used in production. The Environmental Justice Foundation has campaigned against the prevalent use of forced child and adult labor in cotton production in Uzbekistan, the world's third largest cotton exporter. The international production and trade situation has led to "fair trade" cotton clothing and footwear, joining a rapidly growing market for organic clothing, fair fashion or "ethical fashion". The fair trade system was initiated in 2005 with producers from Cameroon, Mali and Senegal.

TRADE

Cotton is bought and sold by investors and price speculators as a tradable commodity on 2 different stock exchanges in the United States of America.

- Cotton futures contracts are traded on the New York Mercantile Exchange (NYMEX) under the ticker symbol TT. They are delivered every year in March, May, July, October, and December.
- Cotton No. 2 futures contracts are traded on the New York Board of Trade (NYBOT) under the ticker symbol CT. They are delivered every year in March, May, July, October, and December.

COTTON GENOME

A public genome sequencing effort of cotton was initiated in 2007 by a consortium of public researchers. They agreed on a strategy to sequence the genome of cultivated, tetraploid cotton. "Tetraploid" means that cultivated cotton actually has two separate genomes within its nucleus, referred to as the A and D genomes. The sequencing consortium first agreed to sequence the D-genome relative of cultivated cotton (*G. raimondii*, a wild Central American cotton species) because of its small size and limited number of repetitive elements. It is nearly one-third the number of bases of tetraploid cotton (AD), and each chromosome is only present once. The A genome of*G. arboreum* would be sequenced next. Its genome is roughly twice the size of *G. raimondii*'s. Part of the difference in size between the two genomes is the amplification of*retrotransposons* (GORGE). Once both diploid genomes are assembled, then research could begin sequencing the actual genomes of cultivated cotton varieties. This strategy is out of necessity; if one were to sequence the tetraploid genome without model diploid genomes, the euchromatic DNA sequences of the AD genomes would co-assemble and the repetitive elements of AD genomes would assembly independently into A and D sequences respectively. Then there would be no way to untangle the mess of AD sequences without comparing them to their diploid counterparts.

The public sector effort continues with the goal to create a high-quality, draft genome sequence from reads generated by all sources. The public-sector effort has generated Sanger reads of BACs, fosmids, and plasmids as well as 454 reads. These later types of reads will be instrumental in assembling an initial draft of the D genome. In 2010, two companies (Monsanto and Illumina), completed enough Illumina sequencing to cover the D genome of *G. raimondii* about 50x. They announced that they would donate their raw reads to the public. This public relations effort gave them some recognition for sequencing the cotton genome. Once the D genome is assembled from all of this raw material, it will undoubtedly assist in the assembly of the AD genomes of cultivated varieties of cotton, but a lot of hard work remains.

MAINTAINING OFFICIAL STANDARDS FOR CLASSIFICATION

To maintain the integrity of the USDA classification system, official standards and standardized procedures have been developed and used throughout the progression of the classification system. Official standards are maintained and provided by the Cotton and Tobacco Program's Standardization and Engineering Division. USDA maintains two basic types of standards for cotton classification: grade standards and instrument standards. For information on purchasing cotton classification standards, contact USDA at cotton.standards@ams.usda.gov.

GRADE STANDARDS

Grade standards are used for manual classification. They specify levels of color and leaf for various grade designations. USDA maintains two types of grade standards: Universal Upland Grade Standards and American Pima Grade Standards. USDA's American Upland cotton standards are referred to as "Universal" standards because they have been adopted by a special governing body and are recognized and used internationally.

USDA has twenty-five official color grades for American Upland cotton and five categories of below-grade color. USDA maintains fifteen physical grade standards for American Upland cotton. Seven of the White color grade standards also serve as official leaf grade standards for American Upland cotton. The remaining grades are descriptive.

For the classification of American Pima cotton, USDA has six official grades for color and leaf, all of which are represented by physical standards. There is also a descriptive standard for cotton that is below grade for color or leaf.

Both Universal Upland and American Pima Grade Standards are valid for only one year, because of gradual changes in color as cotton ages. Grade standards for both American Upland and American Pima cotton are reviewed periodically to ensure that they are still representative of the U.S. crop. If at some point all segments of the U.S. cotton industry agree that the standards are no longer representative of the crop, special measures must be taken to review and amend the standards.

Fig. *USDA maintains fifteen physical Upland Grade Standards and six physical American Pima Grade Standards.*

INSTRUMENT STANDARDS

Instrument standards are cottons used for instrument calibration and verification. These standards include Universal HVI Calibration cotton, Extra-Long Staple (ELS) Calibration cotton, Universal HVI Micronaire Calibration cotton, and Universal HVI Cotton Color and Cotton Trash Standards. These standards serve the USDA and most cotton organizations worldwide as the basis for instrument cotton classification.

Cotton selected for use in instrument calibration must pass rigorous screening procedures. As a first step, USDA conducts an extensive search in the National Database for uniform lots of cotton from the current crop that have fiber properties appropriate for their intended use. Candidate bales are purchased from producers and retested through a rigorous value-establishment process to determine whether they meet the strict certification requirements set for calibration cotton.

ESTABLISHING VALUES FOR CALIBRATION COTTON

In addition to bale uniformity requirements, each bale must meet the length and strength criteria for its intended use. For example, an Upland long/strong calibration cotton bale must have approximate length and strength values of 1.15 to 1.22 inches and 32 to 36 grams per tex, while an Upland short/weak calibration cotton bale must have length and strength values below 1.01 inches and 23 to 26 grams per tex.

Currently, seven laboratories work together to establish values for calibration cottons, including five USDA facilities, one independent laboratory in the U.S. research community, and one well-established international laboratory. The independent U.S. and international laboratories are required to operate under the same rigid specifications as USDA facilities to participate in the value-establishment process.

Cumulatively, the laboratories perform at least 120 tests per bale over a two-day testing period. The results are used to further evaluate uniformity and to determine the values assigned to the calibration cottons. For reference purposes, samples of previously established, or "benchmark," calibration cottons are included in the testing, along with samples from candidate bales. These benchmark cottons provide reference points to assure the continuity of testing levels over time. If test results within a bale are outside of the prescribed limits, the bale is rejected. If all testing criteria are met, the bale is accepted, and its contents are packaged for distribution as calibration cotton.

CLASSIFICATION OF UPLAND COTTON

Measurements for fiber length, length uniformity, fiber strength, micronaire, color grade, trash, and leaf grade are performed by precise High Volume Instruments, in a process commonly referred to as "high volume instrument classification." Only extraneous matter and special conditions are still officially classified by the traditional method of classer determination.

FIBER LENGTH

Fiber length is the average length of the longer half of the fibers (upperhalf mean length). It is reported in both 100ths and 32nds of an inch. Fiber length is measured by passing a "beard" of parallel fibers through an optical sensing point. The beard is formed when fibers from a sample of cotton are automatically grasped by a clamp, then combed and brushed into parallel orientation.

Fiber length is largely influenced by variety, but the cotton plant's exposure to extreme temperatures, water stress, or nutrient deficiencies may result in shorter fibers. Excessive cleaning or drying at the gin may also result in shorter fibers. Fiber length affects yarn strength, yarn evenness, and the efficiency of the spinning process. The fineness of the yarn that can be successfully produced from given fibers also is influenced by fiber length.

LENGTH UNIFORMITY

Length uniformity is the ratio between the mean length and the upperhalf mean length of the fibers, expressed as a percentage. If all of the fibers in the bale were the same length, the mean length and the upperhalf mean length would be the same, and the uniformity would be 100 percent. However, because of natural variation in the length of cotton fibers, length uniformity will always be less than 100 percent.

Length uniformity affects yarn evenness and strength and the efficiency of the spinning process. It is also related to shortfiber content (content of fibers shorter than 1/2 inch). Cotton with a low uniformity index is likely to have a high percentage of short fibers. Such cotton may be difficult to process and is likely to produce low-quality yarn.

FIBER STRENGTH

Strength measurements are reported in grams per tex. A tex unit is equal to the weight in grams of 1,000 meters of fiber. Therefore, the strength reported is the force in grams required to break a bundle of fibers one tex unit in size. Strength measurements are made on the same beards of cotton that are used for measuring fiber length. The beard is clamped in two sets of jaws, 1/8 inch apart, and the amount of force required to break the fibers is determined.

Fig. *Fiber length and strength measurements are made on the same "beard" of cotton.*

Fiber strength is largely determined by variety. However, it may be affected by plant nutrient deficiencies and weather. Fiber strength and yarn strength are highly correlated. Also, cotton with high fiber strength is more likely to withstand breakage during the manufacturing process.

MICRONAIRE

Micronaire is a measure of fiber fineness and maturity. An airflow instrument is used to measure the air permeability of a constant mass of cotton fibers compressed to a fixed volume. The chart below is a guide to interpreting micronaire measurements.

Micronaire can be influenced during the growing period by environmental conditions such as moisture, temperature, sunlight, plant nutrients, and extremes in plant or boll population. Fiber fineness affects processing performance and the quality of the end product in several ways. In the opening, cleaning, and carding processes, low-micronaire or fine-fiber cottons require slower processing speeds to prevent damage to the fibers. Yarns made from finer fiber have more fibers per cross-section, which results in stronger yarns. Dye absorbency and retention are affected by the maturity of the fibers; the greater the maturity, the better the absorbency and retention.

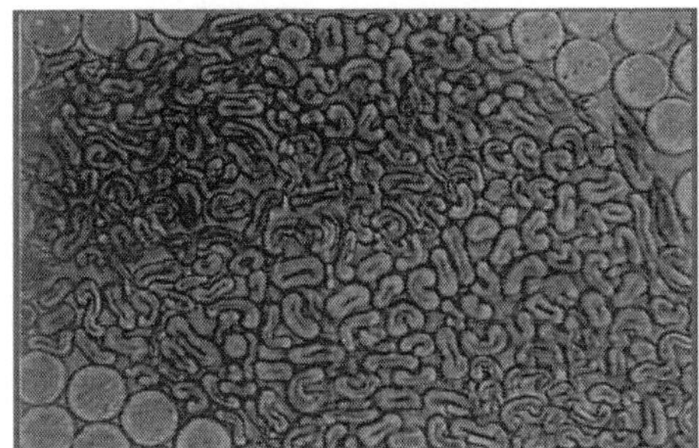
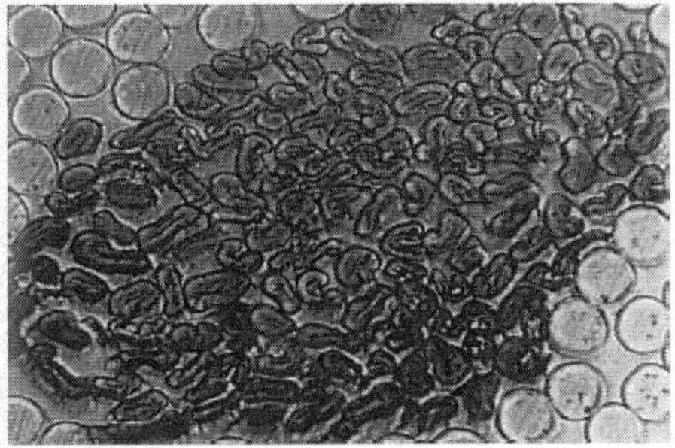

Fig. *Upland cotton with micronaire of 3.8 (left) and 5.2 (right).*

Fig. *The color of cotton is graded by the high volume instrument.*

COLOR GRADE

Color grade is determined by the degree of reflectance (Rd) and yellowness (+b) as established by official standards and measured by the high volume instrument. Reflectance indicates how bright or dull a sample is, and yellowness indicates the degree of pigmentation. A three-digit color code is determined by locating the point at which the Rd and +b values intersect on the color chart for American Upland cotton.

The color of cotton fibers can be affected by rainfall, freezes, insects, fungi, and staining through contact with soil, grass, or cotton-plant leaf. Color can also be affected by excessive moisture and temperature levels during storage, both before and after ginning. Color deterioration because of environmental conditions affects the fibers' ability to absorb and hold dyes and finishes and is likely to reduce processing efficiency.

HVI COLOR GRADES FOR AMERICAN UPLAND COTTON

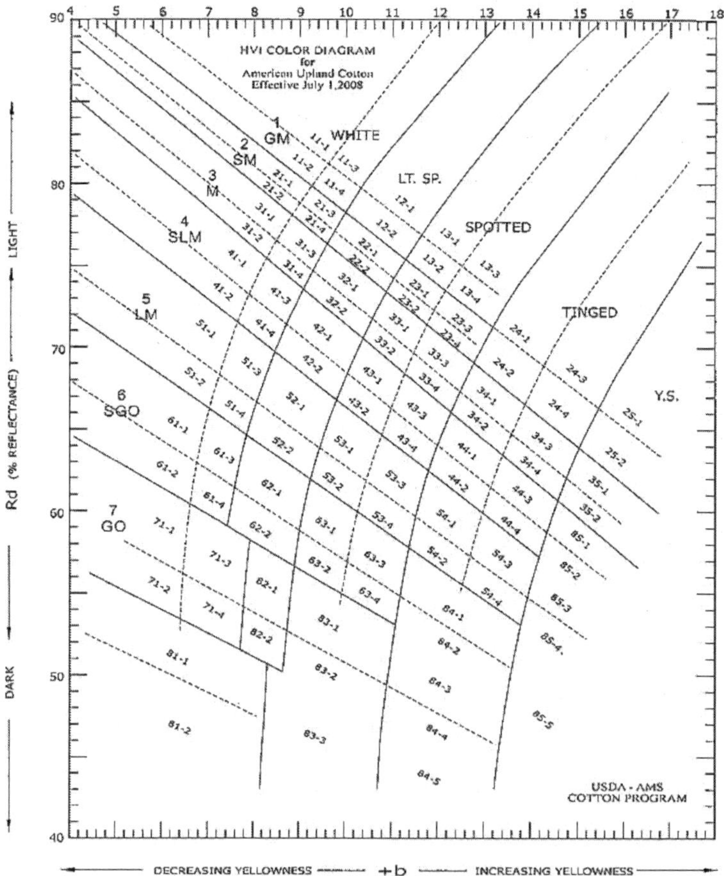

TRASH

Trash is a measure of the amount of non-lint materials in cotton, such as leaf and bark from the cotton plant. The surface of the cotton sample is scanned by a digital camera, and the digital image is analyzed. The percentage of the surface area occupied by trash particles (percent area) and the number of trash particles visible (particle count) are calculated and reported.

The ratio between percent area of trash and trash particle count is a good indicator of the average particle size in a cotton sample. For instance, a low percent area combined with a high particle count indicates a smaller average particle size than does a high percent area with a low particle count.

A high percent area of trash results in greater textile mill processing waste and lower yarn quality. Small trash particles, or "pepper trash," are highly undesirable, because they are more difficult for the mill to remove from the cotton lint than are larger trash particles.

LEAF GRADE

Leaf grade is a measure of the leaf content in cotton. Recent extensive research and development work has resulted in acceptance of instrument leaf grade. Leaf grade is now determined by high volume instrument trash meter percent area and particle count (described above for trash). The leaf grade is calculated from these parameters based on the Universal Upland Grade Standards and American Pima Grade Standards.

Leaf content is affected by plant variety, harvesting methods, and harvesting conditions. The amount of leaf remaining in the lint after ginning depends on the amount present in the cotton before ginning, the amount of cleaning, and the type of cleaning and drying equipment used. Even with the most careful harvesting and ginning methods, a small amount of leaf remains in the cotton lint. From the manufacturing standpoint, leaf content is all waste, and there is a cost factor associated with its removal. Also, small particles cannot always be successfully removed, and these particles may detract from the quality of the finished product.

EXTRANEOUS MATTER

Extraneous matter is any substance in the cotton other than fiber or leaf. Examples of extraneous matter are bark, grass, spindle twist, seedcoat fragments, dust, and oil. The kind of extraneous matter and an indication of the amount (light or heavy) are noted by the classer as a remark on the classification document.

Another factor noted on the classification record under "extraneous matter" is abnormal preparation. "Preparation," or "prep," describes the degree of smoothness or roughness of the ginned cotton lint. Various methods of harvesting, handling, and ginning cotton produce differences in roughness or smoothness of preparation that sometimes are quite apparent. Abnormal preparation of Upland cotton has greatly decreased in recent years as a result of improved harvesting and ginning practices, and now occurs in less than half of one percent of the crop.

MODULE AVERAGING

Module averaging is a voluntary program offered since 1991 to Cotton Program customers at no additional charge. It is a method to improve the reproducibility of the high volume instrument measurements of cotton strength, length, length uniformity, and micronaire.

Improved reproducibility and accuracy enhance the value of U.S. cotton classification and allow all parties to trade U.S. cotton with greater confidence in the quality measurements.

Module averaging does not require a new sampling procedure; it uses the measurements made through the current procedure of obtaining a sample from each side of every bale. With module averaging, all of the individual bale measurements of fiber quality within a module or trailer are averaged, and that average value is assigned to every bale in the module. For example, the individual strength readings for all of the bales in the module are added together and divided by the number of bales in the module to determine the module average for strength, and that value is then assigned as the strength reading for each bale in the module unit. This average serves as the final quality measurement value.

CLASSIFICATION OF AMERICAN PIMA COTTON

Classification procedures for American Pima cotton are similar to those for American Upland cotton, including the use of HVI measurements. The most significant difference is that the American Upland color grade is determined by instrument measurement, while the American Pima color grade is still determined by highly-trained cotton classers. Different grade standards are used because the color of American Pima cotton is a deeper yellow than that of Upland.

Also, the ginning process for American Pima cotton (roller ginned) is not the same as for Upland (saw ginned). The roller gin process results in an appearance that is not as smooth as that obtained with the saw ginned process.

There are six official grades (grades "1" through "6") for American Pima color and six for leaf. All are represented by physical standards. There is a descriptive standard for cotton which is below grade for color or leaf. A different chart is used to convert American Pima fiber length from 100ths to 32nds of an inch. This chart is below.

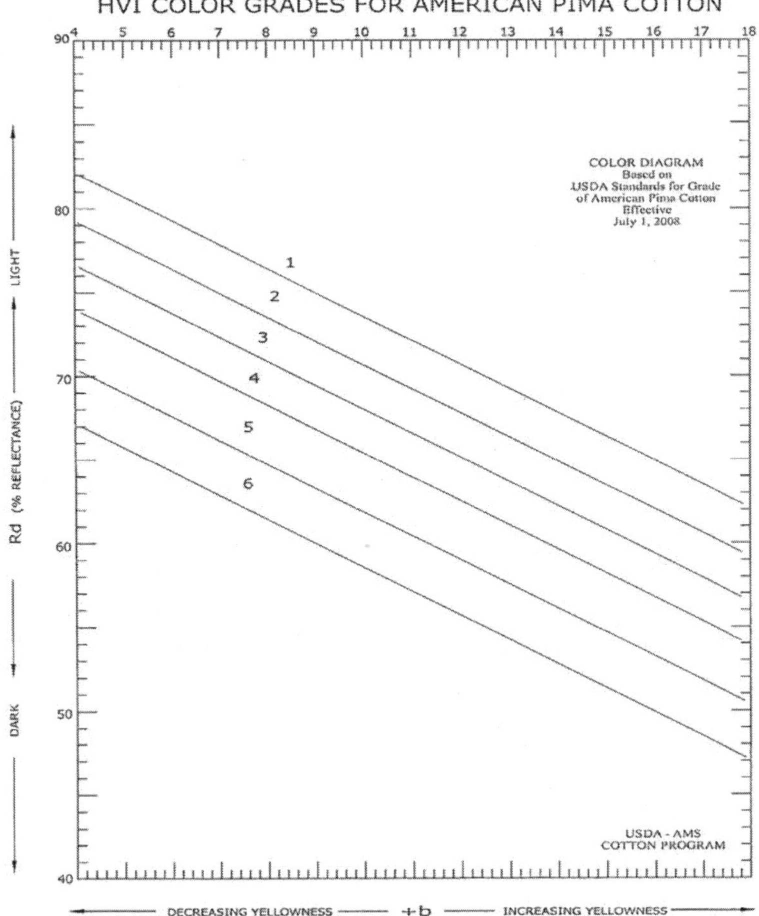

QUALITY AND RELIABILITY OF CLASSIFICATION DATA

Both manual and instrument classification are closely monitored to assure high-quality results. The monitoring of quality is performed mainly through the operations of the Cotton and Tobacco Program's Quality Assurance Division. Several tools and programs are in place to manage quality. These include laboratory conditioning, sample conditioning, equipment performance specifications, instrument calibration, in-house monitoring, and USDA's checklot program.

LABORATORY CONDITIONING

Atmospheric conditions influence the measurement of cotton fiber properties.

Therefore, the temperature and humidity of the classing laboratory must be tightly controlled. Temperature is maintained at 70°F plus or minus 1°F (approximately 21°C plus or minus 1/2°C), and relative humidity is maintained at 65 percent plus or minus 2 percent.

SAMPLE CONDITIONING

Samples are conditioned to bring the moisture content into equilibrium with the approved atmospheric conditions. Conditioned samples will have a moisture content between 6.75 and 8.25 percent (on a dry-weight basis). The conditioned samples are randomly checked to verify that the appropriate moisture content has been reached. Samples may be conditioned passively or actively.

In passive conditioning, the samples are placed in single layers in trays with perforated bottoms to allow free circulation of air.

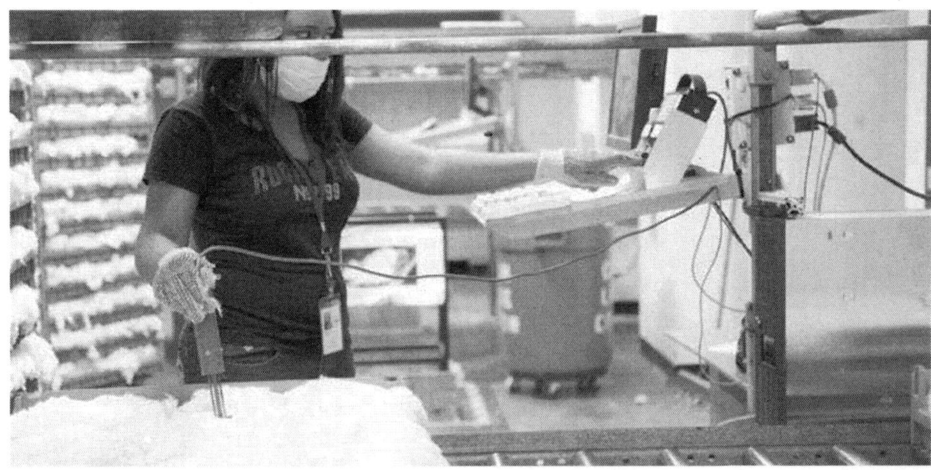

Fig. *Upon arrival at the USDA classing facility, samples are conditioned to standardize moisture content before the classing process begins.*

The samples must be exposed to the approved atmosphere until the specified moisture level is reached, which usually takes at least forty-eight hours.

In active conditioning, a Rapid Conditioning unit is used to draw air set at the approved atmospheric conditions through the sample until the required moisture content for high volume instrument testing is attained. Through active conditioning, the time required to condition samples can be reduced to ten minutes.

EQUIPMENT PERFORMANCE SPECIFICATIONS

It is essential to verify that classing equipment meets minimum performance specifications. "Precision" refers to the ability of an instrument to produce the same measurement result time after time. "Accuracy" refers to how well an instrument measures a certain property in relation to its true value.

Newly purchased equipment must pass a series of thorough tests before being accepted and put into operation. Specifications for the delivery of new equipment include the maximum allowable tolerances for precision shown in the table at right.

Furthermore, all instruments are evaluated annually, typically before each cotton season begins. Testing is done to verify both the precision and the accuracy of instrument measurements.

Calibration of Instruments

Instruments are calibrated for fiber length, length uniformity, micronaire, and fiber strength through the use of calibration cottons. Tiles are used to calibrate color and trash measurements.

IN-HOUSE MONITORING

In addition to calibrating the instruments, each classing office monitors its testing level through an in-house system designed to ensure that the instruments within a classing office provide uniform results. In this system, known-value cottons are tested on each instrument at regular intervals throughout each work shift. If any test value deviates from the known value by more than a specified tolerance, corrective action is taken, such as recalibration or diagnostic testing of the instrument.

CHECKLOT PROGRAM

USDA's checklot program ensures that all USDA classification facilities across the Cotton Belt provide uniform test results. Following instrument testing and classer grading, samples are randomly selected by computer during each work shift, at a rate of approximately one percent of the classing volume. These randomly selected samples are referred to as "checklots." The samples are forwarded by overnight delivery to USDA's Quality Assurance Division, in Memphis, Tennessee, where they are retested. The results are compared with the original classification, and this information is communicated to the originating office, where adjustments are made as necessary. USDA maintains a record of comparisons for each instrument and classer on a daily, weekly, and seasonal basis.

2

History of Cotton

INTRODUCTION

The history of cotton can be traced back to domestication, possibly as far back as 4500 BC. Cotton played an important role in the history of the British Empire, the United States, and India, and continues to be an important crop and commodity.

The history of the domestication of cotton is very complex and is not known exactly. Several isolated civilizations in both the Oldand New World independently domesticated and converted cotton into fabric. All the same tools were invented, including combs, bows, hand spindles, and primitive looms. The oldest cotton textiles were found in graves and city ruins of civilizations from dry climates, where the fabrics did not decay completely. Some of the oldest cotton bolls were discovered in a cave in Tehuacán Valley, Mexico, and were dated to approximately 5500 BCE, but more recent estimates have put the age of these bolls at approximately 3600 BCE. Seeds and cordage dating to about 450BCE have been found in Peru. There is reliable genetic evidence that cotton originated in Peru.

At the source of any plant—in this case wild cotton, the genetic variability is enormously greater and one area of wild cotton in Peru constitutes a "smoking gun.". The Indus Valley civilization started cultivating cotton by 5th or 4th century BCE. By 3000 BCE cotton was being grown and processed in Mexico, and Arizona. Pre-Incan cotton grave cloths were found in Huaca Prieta in Peru, and date back to 2500 BCE, and cotton was mentioned in Hindu hymns in 1500 BCE.

Herodotus, an ancient Greek historian, mentions Indian cotton in the 5th century BCE as "a wool exceeding in beauty and goodness that of sheep." When Alexander the Great invaded India, his troops started wearing cotton clothes that were more comfortable than their previous woolen ones. Strabo, another Greek historian, mentioned the vividness of Indian fabrics, and Arrian told of Indian–Arab trade of cotton fabrics in 130 CE. Egyptians grew and spun cotton from 6–700 CE.

In the 8th century the Muslim conquest of Spain expanded the European cotton trade. By the 15th century, Venice, Antwerp, andHaarlem were important ports for cotton trade, and the sale and transportation of cotton fabrics had become very profitable.

MIDDLE AGES AND THE MODERN ERA

Cotton was a common fabric during the Middle Ages, and was hand-woven on a loom. Cotton manufacture was introduced to Europe during the Muslim conquest of the Iberian Peninsula and Sicily. The knowledge of cotton weaving was spread to northern Italy in the 12th century, when Sicily was conquered by the Normans, and consequently to the rest of Europe. The spinning wheel, introduced to Europe circa 1350, improved the speed of cotton spinning.

Christopher Columbus, in his explorations of the Bahamas and Cuba, found natives wearing cotton ("the costliest and handsomest... cotton mantles and sleeveless shirts embroidered and painted in different designs and colours"), a fact that may have contributed to his incorrect belief that he had landed on the coast of India.

Cotton cloth started to become highly sought-after for the European urban markets during the Renaissance and the Enlightenment. Vasco da Gama, a Portuguese explorer, opened Asian sea trade, which replaced caravans and allowed for heavier cargo. Indian craftspeople had long protected the secret of how to create colourful patterns. However, some converted to Christianity and their secret was revealed by a French Catholic priest, Father Coeurdoux. He revealed the process of creating the fabrics in France, which assisted the European textile industry.

BRITISH EMPIRE

Cotton's rise to global importance came about as a result of the cultural transformation of Europe and Britain's trading empire. Calico and chintz, types of cotton fabrics, became popular in Europe, and by 1664 the East India Company was importing a quarter of a million pieces into Britain. By the 18th century, the middle class had become more concerned with cleanliness and fashion, and there was a demand for easily washable and colourful fabric. Wool continued to dominate the European markets, but cotton prints were introduced to Britain by the East India Company in the 1690s. Imports of calicoes, cheap cotton fabrics from Kozhikode, then known as *Calicut*, in India, found a mass market among the poor. By 1721 these calicoes threatened British manufacturers, and Parliament passed the Calico Act that banned calicoes for clothing or domestic purposes. In 1774 the act was repealed with the invention of machines that allowed for British manufacturers to compete with Eastern fabrics.

Cotton's versatility allowed it to be combined with linen and be made into velvet. It was cheaper than silk and could be imprinted more easily than wool, allowing for patterned dresses for women. It became the standard fashion and, because of its price, was accessible to the general public. New inventions in the 1770s—such as the spinning jenny, the water frame, and the spinning mule—made the British Midlands into a very profitable manufacturing centre. In 1794–1796, British cotton goods accounted for 15.6% of Britain's exports, and in 1804–1806 grew to 42.3%.

The British commercial empire grew the cotton industry enormously. British cotton products were successful in European markets, constituting 40.5% of exports in 1784–1786. Britain's success was also due to its trade with its own colonies, whose settlers maintained British identities, and thus, fashions. With the growth of the cotton industry, manufacturers had to find new sources of raw cotton, and cultivation was expanded to West India. High tariffs against Indian textile workshops, British power in India through the East India Company, and British restrictions on Indian cotton imports transformed India from the source of textiles to a source of raw cotton.Cultivation was also attempted in the Caribbean and West Africa, but these attempts failed due to bad weather and poor soil. The Indian subcontinent was looked to as a possible source of raw cotton, but intra-imperial conflicts and economic rivalries prevented the area from producing the necessary supply.

The Lancashire textile mills were major parts of the British industrial revolution. Their workers had poor working conditions: low wages, child labour, and 18-hour work days.Richard Arkwright created a textile empire by building a factory system powered by water, which was occasionally raided by the Luddites, weavers put out of business by the mechanization of textile production. In the 1790s, James Watt's steam power was applied to textile production, and by 1839 200,000 children worked in Manchester's cotton mills. Karl Marx, who frequently visited Lancashire, may have been influenced by the conditions of workers in these mills in writing *Das Kapital*.

UNITED STATES OF AMERICA

PRE–CIVIL WAR

Anglo-French warfare in the early 1790s restricted access to continental Europe, causing the United States to become an important—and temporarily the largest—consumer for British cotton goods. In 1791, U.S. cotton production was small, at only 900,000 kilograms. Several factors contributed to the growth of the cotton industry in the U.S.: the increasing British demand; the popularity of wearing a cotton flower to symbolize support of the new nation, innovations in spinning, weaving, and steam power; inexpensive land; and a slave labour force. The cotton gin, invented in 1793 by Eli Whitney, enormously grew the American cotton industry, which was previously limited by the speed of manual removal of seeds from the fibre, and helped cotton to surpass tobacco as the primary cash crop of theSouth. By 1801 the annual production of cotton had reached over 22 million kilograms, and by the early 1830s the United States produced the majority of the world's cotton. Cotton also exceeded the value of all other United States exports combined. The need for fertile land conducive to its cultivation lead to the expansion of slavery in the United States and an early 19th-century land rush known asAlabama Fever.

Cultivation of cotton using slaves brought huge profits to the owners of large plantations, making them some of the wealthiest men in the U.S. prior to the Civil War. In the non-slave-owning states, farms rarely grew larger than what could be cultivated by one family due to scarcity of farm workers. In the slave states, owners of farms could buy many slaves and thus cultivate large areas of land. By the 1850s, slaves made up 50% of the population of the main cotton states: South Carolina, Georgia, Alabama, Mississippi, and Louisiana. Slaves were the most important asset in cotton cultivation, and their sale brought profits to slaveowners outside of cotton-cultivating areas. Thus, the cotton industry contributed significantly to the Southern upper class's support of slavery.

CIVIL WAR

"King Cotton", a phrase used by Southern politicians and authors before the Civil War, verbalized their belief that an independent Confederacy would be economically successful, as well as assuring the South's victory if secession from the Union led to war, because of Britain's reliance on the fibre. Senator James Henry Hammond said in 1858:

Without the firing of a gun, without drawing a sword, should they [Northerners] make war upon us [Southerners], we could bring the whole world to our feet. What would happen if no cotton was furnished for three years?... England would topple headlong and carry the whole civilized world with her. No, you dare not make war on cotton! No power on earth dares make war upon it. Cotton is King.

Cotton diplomacy, the idea that cotton would cause Britain and France to intervene in the Civil War, was unsuccessful. It was thought that the Civil War caused the Lancashire Cotton Famine, a period between 1861–1865 of depression in the British cotton industry, by blocking off American raw cotton. Some, however, suggest that the Cotton Famine was mostly due to overproduction and price inflation caused by an expectation of future shortage.

Prior to the Civil War, Lancashire companies issued surveys to find new cotton-growing countries if the Civil War were to occur and reduce American exports. India was deemed to be the country capable of growing the necessary amounts. Indeed, it helped fill the gap during the war, making up only 31% of British cotton imports in 1861, but 90% in 1862 and 67% in 1864. Additionally, the main purchasers of cotton, Britain and France, began to turn to Egyptian cotton. The Egyptian government of Viceroy Isma'il took out substantial loans from European bankers and stock exchanges. After the American Civil War ended in 1865, British and French traders abandoned Egyptian cotton and returned to cheap American exports, sending Egypt into a deficit spiral that led to the country declaring bankruptcy in 1876, a key factor behind Egypt's occupation by the British Empire in 1882.

The South continued to be a one-crop economy until the 20th century, when the New Deal and World War II encouraged diversification. Many ex-slaves as well as poor whites worked in the share-cropping system in serf-like conditions.

MODERN HISTORY

BOLL WEEVILS

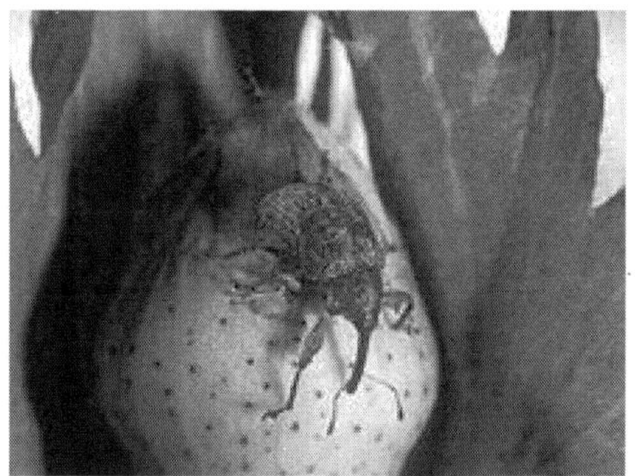

Fig. A boll weevil

" The farmer said to the merchantI need some meat and meal.Get away from here, you son-of-a-gun,You got boll weevils in your field.Going to get your home, going to get your home. "

—/Carl Sandburg's version of "The Boll Weevil Song", 1920

Boll weevils, insects that entered the United States from Mexico in 1892, created 100 years of problems for the U.S. cotton industry. Many consider the boll weevil almost as important as the Civil War as an agent of change in the South, forcing economic and social changes. In total, the boll weevil is estimated to have caused $22 billion in damages. In the late 1950s, the U.S. cotton industry faced economic problems, and eradication of the boll weevil was prioritized. The Agricultural Research Service built the Boll Weevil Research Laboratory, which came up with detection traps and pheromone lures. The program was successful, and pesticide use reduced significantly while the boll weevil was eradicated in some areas.

AFRICA AND INDIA

After the Cotton Famine, the European textile industry looked to new sources of raw cotton. The African colonies of West Africa and Mozambique provided a cheap supply. Taxes and extra-market means again discouraged local textile production. Working conditions were brutal, especially in the Congo, Angola, and Mozambique. Several revolts occurred, and a cotton black market created a local textile industry. In recent history, United States agricultural subsidies have depressed world prices, making it difficult for African farmers to compete.

India's cotton industry struggled in the late 19th century because of unmechanized production and American dominance of raw cotton export. India, ceasing to be a major exporter of cotton goods, became the largest importer of British cotton textiles. Mohandas Gandhi believed that cotton was closely tied to Indian self-determination. In the 1920s he launched the Khadi Movement, a massive boycott of British cotton goods. He urged Indians to use simple homespun cotton textiles, khadi. Cotton became an important symbol in Indian independence. During World War II, shortages created a high demand for khadi, and 16 million yards of cloth were produced in nine months. TheBritish Raj declared khadi subversive; damaging to the British imperial rule. Confiscation, burning of stocks, and jailing of workers resulted, which intensified resistance.In the second half of the 20th century, a downturn in the European cotton industry led to a resurgence of the Indian cotton industry. India began to mechanize and was able to compete in the world market.

DECLINE IN THE BRITISH COTTON INDUSTRY

Fig. British textile mills in 1913

In 1912, the British cotton industry was at its peak, producing eight billion yards of cloth. In World War I, cotton couldn't be exported to foreign markets, and some countries built their own factories, particularly Japan. By 1933 Japan introduced 24-hour cotton production and became the world's largest cotton manufacturer. Demand for British cotton slumped, and during the interwar period 345,000 workers left the industry and 800 mills closed.

India's boycott of British cotton products devastated Lancashire and Blackburn, and 74 mills closed in under four years.

In World War II, the British cotton industry saw an upturn and an increase in workers, with Lancashire mills being tasked with creating parachutes and uniforms for the war.

In the 1950s and '60s, many workers came from the Indian sub-continent and were encouraged to look for work in Lancashire. An increase in the work force allowed mill owners to introduce third (night) shifts. This resurgence in the textile industry did not last long, and by 1958, Britain had become a net importer of cotton cloth.

Modernization of the industry was attempted in 1959 with the Cotton Industry Act. Mill closures occurred in Lancashire, and it was failing to compete with foreign industry. During the 1960s and '70s, a mill closed in Lancashire almost once a week. By the 1980s, the textile industry of North West Britain had almost disappeared.

ECONOMY

Textile mills have moved from Western Europe to, more recently, lower-wage areas. Industrial production is currently mostly located in countries like India, Bangladesh, China, and in Latin America. In these regions labour is much less expensive than in the first world, and attracts poor workers. Biotechnology plays an important role in cotton agriculture as genetically modified cotton that can resist Roundup, a herbicide made by the company Monsanto, as well as repel insects. Organically grown cotton is becoming less prevalent in favour of synthetic fibres made from petroleum products.

The demand for cotton has doubled since the 1980s. The main producer of cotton fibre is now China, at 24%, past the United States at 19% and India at 13%. In 2005/2006, China manufactured 7.15 million tons of textiles, more than double that of India at 3.1 million tons. The leading cotton exporter is the United States, whose production continues to increase due to government subsidies, estimated at $14 billion between 1995 and 2003. The value of cotton lint has been decreasing for sixty years, and the value of cotton has decreased by 50% in 1997–2007. The global textile and clothing industry employs 23.6 million workers, of which 75% are women.

Max Havelaar, a fair trade association, launched a fair trade label for cotton in 2005, the first for a non-food commodity. Working with small producers from Cameroon, Mali, and Senegal, the fair trade agreement increases substantially the price paid for goods and increases adherence to World Labour Organization conventions. A two-year period in Mali has allowed farmers to buy new agricultural supplies and cattle, and enroll their children in school.

COTTON RECYCLING

Fig. Cotton bolls on the cotton plant ready for harvesting and processing into cotton yarn and fabric.

Cotton recycling prevents unneeded wastage and can be a more sustainable alternative to disposal.

PROCESS

Cotton can be recycled from pre-consumer (post-industrial) and post-consumer cotton waste. Pre-consumer waste comes from any excess material produced during the production of yarn, fabrics and textile products, e.g. selvage from weaving and fabric from factory cutting rooms. Post-consumer waste comes from discarded textile products, e.g. used apparel and home textiles. During the recycling process, the cotton waste is first sorted by type and color and then processed through stripping machines that first breaks the yarns and fabric into smaller pieces before pulling them apart into fiber. The mix is carded several times in order to clean and mix the fibers before they are spun into new yarns.

The resulting staple fiber is of shorter length compared to the original fiber length, meaning it is more difficult to spin. Recycled cotton is therefore often blended with virgin cotton fibers to improve yarn strengths. Commonly, not more than 30% recycled cotton content is used in the finished yarn or fabric.

Because waste cotton is often already dyed, re-dyeing may not be necessary. Cotton is an extremely resource-intense crop in terms of water, pesticides and insecticides. This means that using recycled cotton can lead to significant savings of natural resources and reduce pollution from agriculture.

USES

Recycled cotton is often combined with recycled plastic bottles to make clothing and textiles, creating very sustainable, earth-conscious products.

Recycled cotton can also be used in industrial settings as polishing and wiper cloths and can even be made into new, high-quality paper. When reduced to its fibrous state, cotton can be used for applications like seat stuffing or home and automotive insulation. It is also sold as recycled cotton yarn for consumers to create their own items. Additionally, cotton waste can be made into a stronger, more durable paper than traditional wood-pulp based paper, which may contain high concentration of acids. Cotton paper is often used for important documents and also for bank notes since it does not wear off as easily. Cotton waste can also be used to grow mushrooms (particularly the indoor cultivation ofVolvariella volvacea otherwise known as Straw Mushrooms).

Even though recycling cotton cuts down on the harsh process of creating brand new cotton products, it is a natural fiber and is biodegradable, so any cotton fibers that cannot be recycled or used further can be composted and will not take up space in landfills.

ORGANIC COTTON

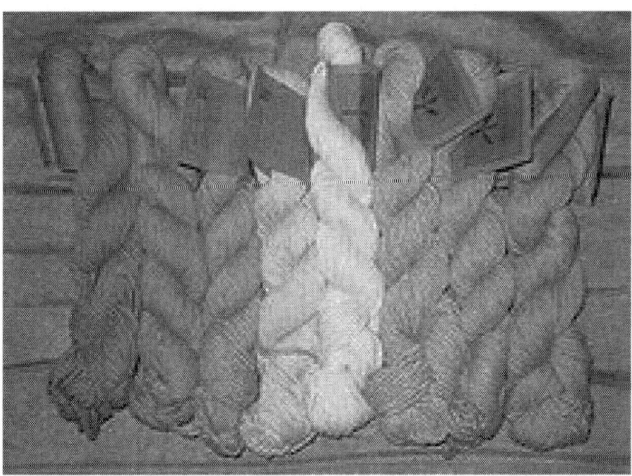

Fig. Organic cotton yarn

Organic cotton is generally understood as cotton and is grown in subtropical countries such as Turkey, China, USA from non genetically modified plants, that is to be grown without the use of any synthetic agricultural chemicals such as fertilizers or pesticides. Its production also promotes and enhances biodiversity and biological cycles. In the United States cotton plantations must also meet the requirements enforced by the National Organic Program (NOP), from the USDA, in order to be considered organic. This institution determines the allowed practices for pest control, growing, fertilizing, and handling of organic crops. As of 2007, 265,517 bales of organic cotton were produced in 24 countries and worldwide production was growing at a rate of more than 50% per year.

ECOLOGICAL FOOTPRINT

Cotton covers 2.5% of the world's cultivated land yet uses 16% of the world's insecticides, more than any other single major crop. Other environmental consequences of the elevated use of chemicals in the non organic cotton growing methods consist of:

- High levels of agrochemicals are used in the production of non-organic, conventional cotton. Cotton production uses more chemicals per unit area than any other crop and accounts in total for 10-16% of the world's pesticides (including herbicides, insecticides, and defoliants).
- Chemicals used in the processing of cotton pollute the air and surface waters.
- Residual chemicals may irritate consumers' skin.
- Decreased biodiversity and shifting equilibrium of ecosystems due to the use of pesticides.

ADVANTAGES

Cotton growers who make the transition to biologically based growing practices expect not only to offer a healthier and cleaner product, but also to benefit the planet. Some of the contributions to the different ecosystems include:

- Protecting surface and groundwater quality (eliminating contaminants in surface runoff)
- Reduced risk in insect and disease control by replacing insecticide with the manipulation of ecosystems
- Long-term prevention of pests through beneficial habitat planting.
- Conservation of biodiversity
- Eliminate the use of toxic chemicals used in cotton

ORGANIC CERTIFICATION

In the USA, it is required by the law that any producer wanting to label and sell a product as "organic" must meet the standards established by the Organic Food Production Act of 1990, enforced by The State organic program (SOP) This act specifies the procedures and regulations for production and handling of organic crops.

ORGANIC SYSTEM

Producers must elaborate an organic production or handling system plan which must also be approved by the state certifying agency or the USDA. This plan must include careful explanation of every process held in the plantation, as well as the frequency with which they are performed. A list of substances used on the crops is also necessary, along with a description of their composition, place where they will be used, and if possible documentation of commercial availability. This inventory of substances is important for the regulation of allowed and prohibited material established by the SOP. Organic cotton growers must also provide a description of the control procedures and physical barriers established to prevent contact of organic and non organic crops on split operations and to avoid contact of organic production with prohibited substance during gestation, harvesting, and handling operations . This production plan can also be transferred to other states as long as it has already been approved by a certifying agency.

PRODUCTION

Production requirements are specifically the set of changes that must be made to field and farming practices in order for a crop to be considered organic. To begin with, organic fields must go through a cleansing period of three years, without the use of any prohibited substances, before planting the first organic crop. Fields must also be equipped with physical barriers and buzzers in order to prevent contact of organic crops with any chemical substance product of surface runoff from crops nearby. Producers must also strive to promote soil fertility through cultivation practices while maintaining or improving the physical, chemical, and biological condition of the soil and minimizes soil erosion. Organic growers must also implement practices to support biodiversity. Such practices include integrated pest management (IPM), which consists of the manipulation of ecosystems that benefit both the crops and the organisms that live around it. In addition to these practices, producers may only apply crop nutrients and soil amendments included on the National List of synthetic substances allowed in crop production.

HANDLING

Handling procedures are all the processes related to product packaging, pest control in handling processing facilities among others. The SOP allows the use of mechanical or biological methods for the purpose of retarding spoilage of products, but at the same time it prohibits the use of volatile synthetic solvents in processed products or any ingredient that is labeled as organic.

PESTICIDES

Since organic cotton is grown without the use of synthetic pesticides, it should contain fewer synthetic pesticides than conventional cotton. Pesticides used in the production of conventional cotton include orthophosphates such as phorate and methamidophos, endosulfan (highly toxic to farmers, but not very environmentally persistent) and aldicarb.Other pesticides persisting in cotton fields in the United States include Trifluralin, Toxaphene and DDT. Although the last two chemicals are no longer used in the United States their long breakdown period and difficulty in removal ensures their persistence. Thus even organic cotton fields may contain them since conventional cotton fields can be transitioned to organic fields in 2–3 years. Instead, organic production allows the use of poisons extracted from plants or animals.

Over time though, studies have been done to find alternatives to conventional pesticide substances. These nonconventional farmers have given up their land and its yields to the testing of different, more organic ways of pest control. Organic farmers argue that conventional farmers don't know the longterm effects of the pesticides they use, especially when the evidence is hidden under the soil. Some farmers in the US use composted tea leaves to act as a substitute for pesticides. Research continues to seek new environmentally friendly ways to rid the soil of harmful pesticides. There has even been a study on using certain animal manure, like chickens, to decrease pest population.

EXPANDING INDUSTRY

Diverse institutions and campaigns are now educating the community about organic cotton and supporting growers on the switch to organic farming. The Sustainable Cotton Project is helping farmers in the transition from chemically dependent crops to more biological sound approaches. This institution has launched the Cleaner Cotton project, which promises to produce cotton with 73% less use of chemicals. In 2003, SCP joined the Community Alliance with Family Farmers (CAFF) to strengthen its operations and reach other farm and consumer audiences. CAFF and SCP provide growers with information about biological farming techniques and educate the public about the importance of reducing chemical use in fiber and food production and supporting local farmers.

REGIONAL

Organic cotton is currently being grown successfully in many countries; the largest producers (as of 2007) are Turkey, India and China.

Organic cotton production in Africa takes place in at least 8 countries. The earliest producer (1990) was the SEKEM organization in Egypt; the farmers involved later convinced the Egyptian government to convert 400,000 hectares of conventional cotton production to integrated methods, achieving a 90% reduction in the use of synthetic pesticides in Egypt and a 30% increase in yields.

Various companies including Nike, Wal-Mart, and C&A (a European Fashion distributor) include or have switched to organic cotton. As of 2011, China, the U.S., India, Pakistan, Brazil, Turkey, Greece, Australia, Syria, Mali, and Egypt are all producing organic cotton. With this rise in demand from 2007 to 2011 more and more countries are making the switch.

THE IMPORTANCE OF COTTON

Today, the world uses more cotton than any other fiber, and cotton is a leading cash crop in the U.S. At the farm level alone, the production of each year's crop involves the purchase of more than $5.3 billion worth of supplies and services. This stimulates business activities for factories and enterprises throughout the country. Processing and handling of cotton after it leaves the farm generates even more business activity. Annual business revenue stimulated by cotton in the U.S. economy exceeds $120 billion, making cotton America's number one value-added crop.

Cotton is a part of our daily lives from the time we dry our faces on a soft cotton towel in the morning until we slide between fresh cotton sheets at night. It has hundreds of uses, from blue jeans to shoe strings. Clothing and household items are the largest uses, but industrial products account from many thousands of bales.

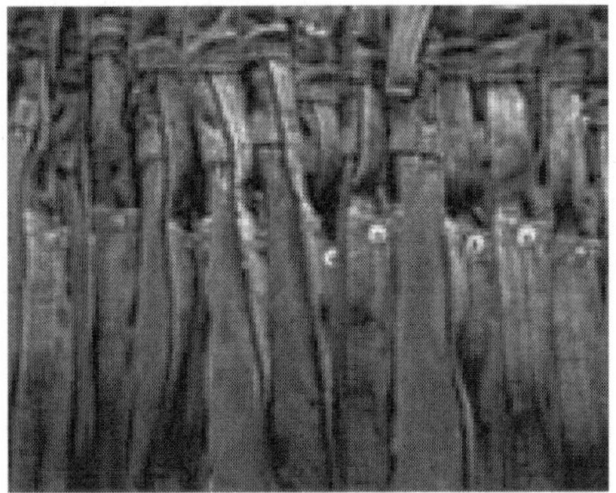

All parts of the cotton plant are useful. The most important is the fiber or lint, which is used in making cotton cloth. Linters – the short fuzz on the seed – provide cellulose for making plastics, explosives and other products. Linters also are incorporated into high quality paper products and processed into batting for padding mattresses, furniture and automobile cushions.

The cottonseed is crushed in order to separate its three products – oil, meal and hulls. Cottonseed oil is used primarily for shortening, cooking oil and salad dressing. The meal and hulls that remain are used either separately or in combination as livestock, poultry and fish feed and as fertilizer. The stalks and leaves of the cotton plant are plowed under to enrich the soil.

Some cottonseed also is used as high-protein concentrate in baked goods and other food products.

WHERE COTTON GROWS

Cotton Cotton grows in warm climates and most of the world's cotton is grown in the U.S., Uzbekistan, the People's Republic of China and India. Other leading cotton-growing countries are Brazil, Pakistan and Turkey.

In this country, the major cotton-producing states are: Alabama, Arizona, Arkansas, California, Georgia, Louisiana, Mississippi, Missouri, New Mexico, North Carolina, Oklahoma, South Carolina, Tennessee and Texas, Florida, Kansas and Virginia.

The yield in the U.S. averages approximately 1 1/3 bales per acres and about 1,078 pounds of seed. A U.S. bale weighs around 500 pounds. This yield is about twice as much as in 1950 and is due to better land use, improved plant varieties, mechanization, fertilization and irrigation. It also is a result of much better control of disease, weeds and insects. A major part of the credit for this progress goes to scientists working at experiment stations and in laboratories, and to agricultural extension workers who bring the findings to farmers.

HOW COTTON IS GROWN

After cotton has been harvested, producers who use conventional tillage practices cut down and chop the cotton stalks. The next step is to turn the remaining residue underneath the soil surface. Producers who practice a style of farming called conservation tillage often choose to leave their stalks standing and leave the plant residue on the surface of the soil. In the spring, farmers prepare for planting in several ways. Producers who plant using no-till or conservation tillage methods, use special equipment designed to plant the seed through the litter that covers the soil surface. Producers who employ conventional tillage practices, plow or "list" the land into rows forming firm seed-beds for planting. Producers in south Texas plant cotton as early as February. In Missouri and other northern parts of the Cotton Belt, they plant as late as June.

Seeding is done with mechanical planters which cover as many as 10 to 24 rows at a time. The planter opens a small trench or furrow in each row, drops in the right amount of seed, covers them and packs the earth on top of them. The seed is planted at uniform intervals in either small clumps ("hill-dropped") or singularly ("drilled"). Machines called cultivators are used to uproot weeds and grass, which compete with thecotton plant for soil nutrients, sunlight and water.

About two months after planting, flower buds called squares appear on the cotton plants. In another three weeks, the blossoms open. Their petals change from creamy white to yellow, then pink and finally, dark red. After three days, they wither and fall, leaving green pods which are called cotton bolls.

Inside the boll, which is shaped like a tiny football, moist fibers grow and push out from the newly formed seeds. As the boll ripens, it turns brown. The fibers continue to expand under the warm sun. Finally, they split the boll apart and the fluffy cotton bursts forth. It looks like white cotton candy.

Since hand labor is no longer used in the U.S. to harvest cotton, the crop is harvested by machines, either a picker or a stripper. Cotton picking machines have spindles that pick (twist) the seed cotton from the burrs that are attached to plants' stems. Doffers then remove the seed cotton from the spindles and knock the seed cotton into the conveying system.

Conventional cotton stripping machines use rollers equipped with alternating bats and brushes to knock the open bolls from the plants into a conveyor.

A second kind of stripper harvester uses a broadcast attachment that looks similar to a grain header on a combine. All harvesting systems use air to convey and elevate the seed cotton into a storage bin referred to as a basket. Once the basket is full, the stored seed cotton is dumped into a boll buggy, trailer or module builder.

HOW COTTON IS GINNED AND MARKETED

Today, nearly all cotton is stored in modules, which look like giant loaves of bread. Modules allow the cotton to be stored without loosing yield or quality prior to ginning. Specially designed trucks pick up modules of seed cotton from the field and move them to the gin. Modern gins place modules in front of machines called module feeders. Some module feeders have stationary heads, in which case, giant conveyors move the modules into the module feeder. Other module feeders are self-propelled and move down a track that along side the modules. The module feeders literally break the modules apart and "feed" the seed cotton into the gin.

Other gins use powerful pipes to suck the cotton into the gin building. Once in the cotton gin, the seed cotton moves through dryers and through cleaning machines that remove the gin waste such as burs, dirt, stems and leaf material from the cotton. Then it goes to the gin stand where circular saws with small, sharp teeth pluck the fiber from the seed.

From the gin, fiber and seed go different ways. The ginned fiber, now called lint, is pressed together and made into dense bales weighting about 500 pounds. To determine the value of cotton, samples are taken from each bale and classed according to fiber length (staple), strength, micronaire, color and cleanness. Producers usually sell their cotton to a local buyer or merchant who, in turn, sells it to a textile mill either in the United States or a foreign country.

The seed usually is sold by the producer to the gin. The ginner either sells for feed or to an oil mill where the linters (downy fuzz) are removed in an operation very much like ginning. Linters are baled and sold to the paper, batting and plastics industries, while the seed is processed into cottonseed oil, meal and hulls.

HOW COTTON IS SPUN AND WOVEN

At the textile mill, the bales are opened by machines, and the lint is mixed and cleaned further by blowing and beating. The short lint that comes out usually is separated and sold for use in other industries. The best part of the lint consists of fibers about 1 inch to 1 ¾ inches long.

The mixed and fluffed-up cotton goes into a carding machine which cleans the fibers some more and makes them lie side by side. The combing action of the carding machine finishes the job of cleaning and straightening the fibers, and makes them into a soft, untwisted rope called a sliver (pronounced sly-ver).

On modern spinning frames, yarn is mare directly from the sliver. The spinning devices take fibers from the sliver and rotate it up to 2,500 revolutions in a second twist that makes fibers into a yarn for weaving or knitting into fabrics.

Machines called looms weave cotton yarns into fabrics the same way the first handweaving frames did. Modern looms work at great speeds, interlacing the length-wise yarns (warp) and the crosswise yarns (filling). The woven fabric, called gray goods, is sent to a finishing plant where it is bleached, pre-shrunk, dyed, printed and given a special finish before being made into clothing or products for the home. Other machines make knits for use in shirts, sweaters or blankets.

This, then, is the story of cotton – where and how it is grown, marketed, processed and manufactured into the many useful products that have served the world so well for so long. It is a never-ending story, as scientists continue to develop better ways to produce and use one of the world's oldest fibers – cotton.

3

Cotton Fibers

INTRODUCTION

Cotton today is the most used textile fiber in the world. Its current market share is 56 percent for all fibers used for apparel and home furnishings and sold in the U.S. . Another contribution is attributed to nonwoven textiles and personal care items. It is generally recognized that most consumers prefer cotton personal care items to those containing synthetic fibers. World textile fiber consumption in 1998 was approximately 45 million tons. Of this total, cotton represented approximately 20 million tons. . The earliest evidence of using cotton is from India and the date assigned to this fabric is 3000 B.C. There were also excavations of cotton fabrics of comparable age in Southern America. Cotton cultivation first spread from India to Egypt, China and the South Pacific. Even though cotton fiber had been known already in Southern America, the large-scale cotton cultivation in Northern America began in the 16th century with the arrival of colonists to southern parts of today's United States. .The largest rise in cotton production is connected with the invention of the saw-tooth cotton gin by Eli Whitney in 1793. With this new technology, it was possible to produce more cotton fiber, which resulted in big changes in the spinning and weaving industry, especially in England.

COTTON CONSUMPTION AND PRODUCTION IN MILLION TONS IN YEAR 2002

The graph bellow shows Production and consumption of leading cotton producing countries in Millions of tones in year 2002 .

COUNTRIES	PRODUCTION	CONSUMPTION
US	3.8	1.7
India	2.5	3
Pakistan	1.8	1.9
Turkey	0.9	1.4
Brazil	0.7	0.9
Indonesia	0.4	0.6
China	4.8	5.9

CHARATERISTICS OF COTTON

Cotton, as a natural cellulosic fiber, has a lot of characteristics, such as;

- Comfortable Soft hand
- Good absorbency
- Color retention
- Prints well
- Machine-washable
- Dry-cleanable
- Good strength
- Drapes well
- Easy to handle and sew

END USES OF COTTON:

- Apparel - Wide range of wearing apparel: blouses, shirts, dresses, childrenswear, active wear, separates, swimwear, suits, jackets, skirts, pants, sweaters, hosiery, neckwear.
- Home Fashion - curtains, draperies, bedspreads, comforters, throws, sheets, towels, table cloths, table mats, napkins

STRUCTURE AND PROPERTIES OF COTTON FIBERS

FIBER STRUCTURE AND FORMATION

The botanical name of American Upland cotton is *Gossypium Hirsutum* and has been developed from cottons of Central America. Upland varieties represent approximately 97% of U.S. production .

Each cotton fiber is composed of concentric layers. The cuticle layer on the fiber itself is separable from the fiber and consists of wax and pectin materials. The primary wall, the most peripheral layer of the fiber, is composed of cellulosic crystalline fibrils. The secondary wall of the fiber consists of three distinct layers. All three layers of the secondary wall include closely packed parallel fibrils with spiral winding of 25-35° and represent the majority of cellulose within the fiber. The innermost part of cotton fiber- the lumen- is composed of the remains of the cell contents. Before boll opening, the lumen is filled with liquid containing the cell nucleus and protoplasm. The twists and convolutions of the dried fiber are due to the removal of this liquid. The cross section of the fiber is bean-shaped, swelling almost round when moisture absorption takes place.

The overall contents are broken down into the following components.

RAW COTTON COMPONENTS:

During scouring (treatment of the fiber with caustic soda), natural waxes and fats in the fiber are saponified and pectin's and other non-cellulose materials are released, so that the impurities can be removed by just rinsing away. After scouring, a bleaching solution (consisting of a stabilized oxidizing agent) interacts with the fiber and the natural color is removed. Bleaching takes place at elevated temperature for a fixed period of time . Mercerization is another process of improving sorption properties of cotton. Cotton fiber is immersed into 18- 25% solution of sodium hydroxide often under tension . The fiber obtains better luster and sorption during mercerization.

80-90%	Cellulose
6-8%	Water
0.5 - 1%	Waxes and fats
0 - 1.5%	Proteins
4 - 6%	Hemicelluloses and pectin's
1 - 1.8%	Ash

After scouring and bleaching, the fiber is 99% cellulose. Cellulose is a polymer consisting of anhydroglucose units connected with 1,4 oxygen bridges in the beta position. The hydroxyl groups on the cellulose units enable hydrogen bonding between two adjacent polymer chains. The degree of polymerization of cotton is 9,000-15,000 . Cellulose shows approximately 66% crystallinity, which can be determined by X-ray diffraction, infrared spectroscopy and density methods.

Each crystal unit consists of five chains of anhydroglucose units, parallel to the fibril axis. One chain is located at each of the corners of the cell and one runs through the center of the cell. The dimensions of the cell are a = 0.835nm, b = 1.03 nm and c = 0.79 nm. The angle between ab and BC planes is 84° for normal cellulose, i.e., Cellulose I .

REPEAT UNIT OF CELLULOSE

The current consensus regarding cellulose crystallinity (X-ray diffraction) is that fibers are essentially 100% crystalline and that very small crystalline units imperfectly packed together cause the observed disorder.

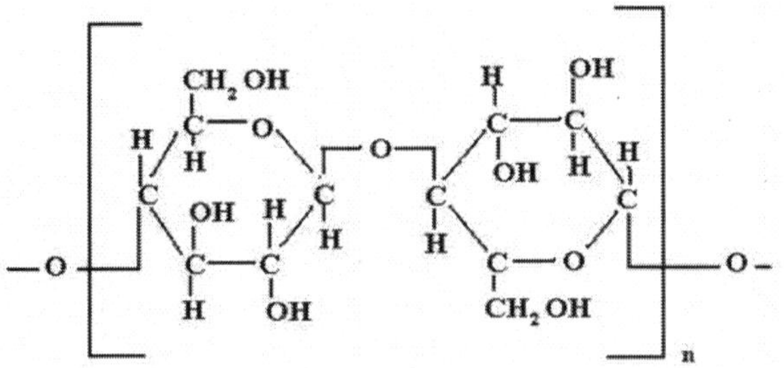

The density method used to determine cellulose crystallinity is based on the density gradient column, where two solvents of different densities are partially mixed. Degree of Crystallinity is, then, determined from the density of the sample, while densities of crystalline and amorphous cellulose forms are known (1.505 and 1.556 respectively). Orientation of untreated cotton fiber is poor because the crystallites are contained in the micro fibrils of the secondary wall, oriented in the steep spiral (25-30°) to the fiber axis.

PHYSICAL PROPERTIES OF COTTON

FIBER LENGTH

Fiber length is described as "the average length of the longer one-half of the fibers (upper half mean length)" This measure is taken by scanning a "beard " of parallel fibers through a sensing region. The beard is formed from the fibers taken from the sample, clasped in a holding clamp and combed to align the fibers. Typical lengths of Upland cottons might range from 0.79 to 1.36in.

Cottons come from the cotton plant; the longer strand types such as Pima or Sea Island produce the finest types of cotton fabrics .

LENGTH UNIFORMITY

Length uniformity or uniformity ratio is determined as " a ratio between the mean length and the upper half mean length of the fibers and is expressed as a percentage". Typical comparisons are illustrated below.

LENGTH UNIFORMITY	UNIFORMITY INDEX [%]
Very High	>85
High	83-85
Intermediate	80-82
Low	77-79
Very Low	<77

Low uniformity index shows that there might be a high content of short fibers, which lowers the quality of the future textile product.

FIBER STRENGTH

Fiber strength is measured in grams per denier. It is determined as the force necessary to break the beard of fibers, clamped in two sets of jaws, (1/8 inch apart) . Typical tensile levels are illustrated. The breaking strength of cotton is about 3.0~4.9 g/denier, and the breaking elongation is about 8~10%.

DEGREE OF STRENGTH	FIBER STRENGTH [g/tex]
Very Strong	>31
Strong	29-30
Average	26-28
Intermediate	24-25
Weak	<23

MICRONAIRE

Micronaire measurements reflect fiber fineness and maturity. A constant mass (2.34 grams) of cotton fibers is compressed into a space of known volume and air permeability measurements of this compressed sample are taken. These, when converted to appropriate number, denote Micronaire values.

COTTON RANGE	MICRONAIRE READING
Premium	3.7-4.2
Base Range	4.3-4.9
Discount Range	>5.0

COLOR

The color of cotton samples is determined from two parameters: degree of reflectance (Rd) and yellowness (+b). Degree of reflectance shows the brightness of the sample and yellowness depicts the degree of cotton pigmentation. A defined area located in a Nickerson-Hunter cotton colorimeter diagram represents each color code. The color of the fibers is affected by climatic conditions, impact of insects and fungi, type of soil, storage conditions etc. There is five recognized groups of color: white, gray, spotted, tinged, and yellow stained. As the color of cotton deteriorates, the process ability of the fibers decreases. Work at the University of Tennessee has led to color measurement using both a spectrometer CIE-based average color measurement and a color uniformity measurement using image analysis to improve the accuracy and provide additional measurement for color grading . Later the investigators developed two color grading systems using expert system and neural networks.

TRASH

A trash measurement describes the amount of non-lint materials (such as parts of cotton plant) in the fiber. Trash content is assessed from scanning the cotton sample surface with a video camera and calculating the percentage of the surface area occupied by trash particles. The values of trash content should be within the range from 0 to 1.6%. Trash content is highly correlated to leaf grade of the sample.

LEAF GRADE

Leaf grade is provided visually as the amount of cotton plant particles within the sample. There are seven leaf grades (#1-#7) and one below grade (#8).

PREPARATION

Preparation is the classer's interpretation of fiber process ability in terms of degree of roughness or smoothness of ginned cotton.

EXTRANEOUS MATTER

Extraneous matter is all the material in the sample other than fiber and leaf. The classer either as "light" or "heavy" determines the degree of extraneous matter.

NEPS

A nep is a small tangled fiber knot often caused by processing. Neps can be measured by the AFIS nep tester and reported as the total number of neps per 0.5 grams of the fiber and average size in millimeters. Nep formation reflects the mechanical processing stage, especially from the point of view of the quality and condition of the machinery used.

CHEMICAL PROPERTIES OF COTTON

Cotton swells in a high humidity environment, in water and in concentrated solutions of certain acids, salts and bases. The swelling effect is usually attributed to the sorption of highly hydrated ions. The moisture regain for cotton is about 7.1~8.5% and the moisture absorption is 7~8%. Cotton is attacked by hot dilute or cold concentrated acid solutions. Acid hydrolysis of cellulose produces hydro-celluloses. Cold weak acids do not affect it. The fibers show excellent resistance to alkalis. There are a few other solvents that will dissolve cotton completely.

One of them is a copper complex of cupramonium hydroxide and cupriethylene diamine (Schweitzer's reagent) Cotton degradation is usually attributed to oxidation, hydrolysis or both. Oxidation of cellulose can lead to two types of so-called oxy-cellulose , depending on the environment, in which the oxidation takes place.

INSERT FORMULA OR EQUATION: OXY-CELLULOSE

Also, cotton can degrade by exposure to visible and ultraviolet light, especially in the presence of high temperatures around 250~397° C and humidity. Cotton fibers are extremely susceptible to any biological degradation (microorganisms, fungi etc.)

OPTICAL PROPERTIES OF COTTON

Cotton fibers show double refraction when observed in polarized light. Even though various effects can be observed, second order yellow and second order blue is characteristic colors of cellulosic fibers. A typical birefringence value as shown in the table of physical properties, is 0.047.

COTTON CLASSIFICATION

Cotton classification is used to determine the quality of the cotton fiber in terms of grade, length and Micronaire . USDA classification specifically identifies the characteristics of fiber length, length uniformity, strength, Micronaire, color, preparation, leaf and extraneous matter. In the past, these qualities were classified just by hand-and-eye of an experienced classer. Since 1991, all classification has been carried out with a set of up-to-date instruments, called "HVI"(High Volume Instrumentation) classification . However, measuring techniques of other qualities of cotton fiber, such as fiber maturity and short fiber content, are also being developed.

COTTON IN NON-WOVENS

Cotton is the most important apparel fiber throughout the world. It is a fiber that was used fairly extensively during the early, developmental period of the Nonwovens business primarily because the emerging dry-laid producers came from the textile industry and had an intimate knowledge of cotton and its processing characteristics .It was in the early part of 20th Century that a few cotton mills in the US wanted to find ways to upgrade the waste cotton fibers into saleable products. The first method used was bonding the short cotton fibers (fiber waste) with latex and resin.

These products were used mainly as industrial wipes. After World War II, products like draperies, tablecloths, napkins and wiping towels were developed. It was realized that woven fabrics have much better properties than Nonwovens; so, the approach was to claim the market where superior qualities of woven or knit fabrics were not essential but where qualities better than those of paper were needed. As the quality requirements for nonwoven fabrics increased and particularly as the need for white, clean fabric emerged; the use of raw cotton became unacceptable and was abandoned by the industry except for a few isolated product areas. Within the last decade, bleached cotton fiber suitable for processing on conventional nonwoven equipment has become available and has substantially increased interest in this fiber. This is particularly true in medical and healthcare applications, wiping and wiper markets, and some apparel markets.

The raw cotton consists of about 96% cellulose and 4% of waxes, pectin, and other pertinacious and plant material. These minor constituents that must be removed in the scouring and bleaching process to give the soft, clean, white, absorbent fiber that is satisfactory for the nonwovens industry after the application of an appropriate finishing oil. The fiber length of cotton is important, particularly as to its process ability.

Longer staple cotton (0.75 in. to 1.25 in.) is satisfactory for nonwoven production. The fiber has excellent absorbency and feels comfortable against the skin. It has fairly good strength both wet and dry, and has moderate dimensional stability and elastic recovery. But the resilience of cotton is relatively low, unless it is cross-linked by a chemical treatment. In nonwoven applications, the purity and absorbency of bleached cotton are utilized in growing medical and healthcare applications. The spun lace process usually produces such fabrics. For similar reasons, cotton spun lace fabrics are well accepted in personal and related wipes, especially in Japan and the ASIAN region. In a sense, bleached cotton fiber for nonwoven application is a relatively new fiber. It is a comparatively expensive fiber and available from only a few sources.

Consequently, its use still is restricted to specialized applications. This situation is likely to change in the future as the price is further reduced and availability increased.

FIBER PROCESSING

About 30% of world cotton machines harvest production. Australia, Israel and USA are the only countries where all cottons are picked by machines. Fifteen percent of world cotton production is ginned on roller gins and almost all rest of cotton is saw ginned in most countries .Cotton fibers in non-wovens are generally used in their bleached form.

A lot of research and development has taken place for the efficient production of bleached fibers.

The Kier bleaching process produces most of the bleached cotton fibers. Since cotton of lesser grades is useful for non-wovens, a conventional cleaning system does not suffice. This might include a coarse wire carding, called Cotton Master Cleaners, for cleaning the cotton.

- The conventional bleaching method for cottons meant for non-wovens is a 9 step process are:
  a. Fiber opening and cleaning
  b. Alkali scouring application
  c. Alkali reaction stage
  d. Rinsing
  e. Bleach application
  f. Bleach reaction stage
  g. Rinsing
  h. Finish application
  i. Drying

A continuous textile processing system and method have been disclosed recently for producing a nonwoven web containing bleached cotton fibers in a single line system which includes a supply of fibers such as a bale opening device, The final nonwoven web consisting of bleached cotton fibers may be made into highly purified and absorbent wipes, pads, and other articles for medical, industrial, or domestic use .

Finally, there is opening and bale formation.

- Cotton Incorporated patented a processing line, which promised better productivity and quality. It consists of:
  a) Fiber opening and Cleaning
  b) Formation of web
  c) Steam purging and Alkali impregnation onto the sandwiched cotton web between 2 porous conveyors.
  d) After reaction, a pressure squeezing operation.
  e) Similar processes for bleaching and then finishing.

- The recent system for scouring a bleaching of cotton fiber is the Continuous Wet Finishing Technique' patented by Lawrence Girard and Walter E Meyer and assigned to Greenville Machinery Corporation. It consists of:
  a. Opening and Cleaning
  b. Conversion of fibers into a bat, weighing 10-30 ounces/sq. yard, by Needle punching or Air-lay technique.
  c. Scouring
  d. Bleaching
  e. Finishing
  f. Washing
  g. Drying
  h. Fiber opening

Advantages of Continuous Finishing Techniques are:

a) Uniformity of scouring and bleaching
b) Uniformity of finish application
c) Shorter time in process for the materials
d) Lower water consumption and less effluent for treatment
e) The ability to provide additional chemical treatments to the cotton.

. COST OF PRODUCING COTTON

The international cotton advisory committee (ICAC) undertakes a survey of the cost of the production of cotton every three years based on the data from 31 countries. Several factors are considered, such as land rent, fertilizers, insect control, irrigation, harvesting and ginning. The cost of seed cotton is more than $500 in USA to produce one hectare of seed cotton. The net cost of producing lint from one hectare (the value of seed and land rent were excluded from the total cost) is highest in Australia (US$1,056) followed by the USA (US$889), Pakistan (US$814), Zimbabwe (US$426) and China (US$416). It is most expensive to produce a kilogram of lint in the USA (US$1.20), Australia (US$0.75) and china (US$0.48).

WEB PROCESSING WITH COTTON

Cotton fibers are used in the manufacture of nonwovens either alone or in a blend. The various processes for the manufacture of non-wovens are:

HYDROENTANGLEMENT:

This method of bonding provides strength to the Nonwovens, comparable to woven fabric of the same basis weight. This method yields high strength without interfering with the absorbency, tensile strength and aesthetic properties of cotton. This type of nonwovens can be wet processed like the conventional woven textiles for bleaching, dyeing and finishing. To manufacture soft loose nonwovens, partially entangled webs are produced by subjecting cotton webs to low water jet pressures (approx. 300-500 psi). These types of webs can be wet processed in a pad/batch state. The limitations of this process are that production has been limited to fiber blends because of problems in recycling water and the quality of bleached cotton.

NEEDLE PUNCHING:

Needle punched cotton provides highly efficient filter media based on the irregular fiber shape and absorption properties. Increased tenacity in the wet condition can be an important advantage for cotton filters. To build strength, scrim materials can be used as in bed blankets and industrial fabrics. Needles of 36-42 gauges have been found appropriate for the production of cotton needle punched nonwovens. For very heavy fabrics, use is made of gauge 32 and for finer fabrics 40-42 gauge needles are being used.

THERMAL BONDING:

In this process cotton webs with blends of thermoplastic fibers are passed between 2 hot rollers (Calendar rollers). The thermoplastic fiber softens/melts and bonds the web. The initial work was done with polyester as the thermoplastic fiber. Later polypropylene was extended for the study because of economics, density and melting temperature considerations. This was mainly to study the application as a diaper lining material. Substantial work is still being done to develop this type of nonwovens.

OTHER BONDING SYSTEMS:

a. Impregnating the web with a resin or other adhesive material.
b. Stripping off of the web with adhesive, which bonds the fibers together at regular intervals.
c. Stitch bonding: cotton web is stitched like in sewing and the product performance depends on web weight, stitch/inch and type of sewing thread.

APPLICATIONS AND MANUFACTURERS OF COTTON NON-WOVENS

Cotton nonwovens are used as swabs, puffs, wipes, filters, weddings, personal care products like in diapers & feminine hygiene products, semi-durable segments like bedding, household furnishing, pillow fillers, etc.

MANUFACTURERS OF COTTON

- Barhardt manufacturing
- Bba nonwovens veratec
- Brannoc fibers ltd.
- Cotton incorporated
- Ihsan sons (pvt) limited

- Leigh fibers
- Textiles and nonwovens directory

RECENT RESEARCH

- New instrumentation to measure cotton contamination .
- Cotton linters to replace the traditional 100% wood pulp fibers for producing absorbent cores for disposable diapers and famine pads
- New quality measurements of small sample cotton are being developed
- Cotton is being blended with kenaf fibers to improve the softness and hand
- Buckeye Technologies has developed 100% natural cotton for tampon manufacture
- Clustering analysis is developed for cotton trash classification
- New method to improve the dyeability of cotton with reactive dyes.

RECENT DOVELOPMENTS IN COTTON

COLORED COTTON

Cotton fiber is dyed with chemical dyes in order to get wide range of colors. These chemical dyes and their finishing demands large amount of water in turn when these water is disposed they cause soil and water pollution. Many dyes are of chemical origin; particularly the azure ones and these are not environment friendly. Hence many countries, including India, have prohibited use of these dyes.

The negative effects of dyeing can be reduced by naturally colored cotton. This colored cotton is developed by gene transplantation. Crossing the genes from wild cotton varieties with the cultivated white ones develops this colored cotton. The research is being conducted at The University of Agricultural Sciences (UAS), Dharwad Karnataka India, to promote the cultivation of natural colored cotton. The colors that have been developed are White, Orange, Red, Yellow, Green, Purple, Brown, Blue, And Black. These negative effects of dyeing can be avoided by extensive research and growth of colored cotton. (33).

BT COTTON

Cotton requires severe pesticide in order to combat numerous pests after some years of use of pesticide by farmers these pests develop resistance to Particular pesticide. This resistance force farmers to use more amounts of pesticides. BT Cotton is developed by transgenic technique of implanting Bacillus Thuringiens bacterial gene in to cottonseeds, which makes the cotton plant and seeds resistant to majority of pests including bollworm (A. Lepidoptera), Tobacco budworm (Heliothis virescens). Bt cotton is now one of the most widely used transgenic crops. It is currently grown throughout the United States. More than 2 million acres of Bt cotton are grown in the United States alone. Other countries include China, India, and Australia. (34) According to Dept. of Agricultural and Resource Economics, University of Arizona. *Bt* cotton planted from 1996 to 1998 is estimated to have yielded 5% more on average than if traditional and decreased the quantity of foliar spray .

COTTON'S FUTURE TRENDS

The world's cotton fiber production is approximately 89 million bales . In 1997, a production forecast shows that the U.S. is the largest cotton producer (18.4 million bales), followed by China (17.5 million bales), India (12.8 million bales), Pakistan (8.0 million bales) and the former U. S. S. R. republics (7.7 million bales). Other important cotton producers are Australia, Egypt, Turkey, Brazil, Argentina, Paraguay, Greece and Mexico. The highest cotton consumption is attributed to China (21.2 million bales), India (12.9 million bales) and U.S. (11.3 million bales).

SUPPLIES: The world production will increase a little bit. The 1998 U.S cotton crop is best described as a disaster due to cool wet spring in the west and inadequate rainfall in the southeast .

CONSUMPTION: World cotton consumption is lagging a bit behind production. After a surge in the mid-1980s, world cotton consumption has been rather flat. But the long term potential for cotton demand remains large .

All cotton plantings for 1999 are expected to total 14.6 million acres, 9 percent above 1998, and 5 percent greater than 1997. Upland cotton is expected to total 14.2 million acres, up 9 percent from last year. Growers planted 318,200 acres of American-Pima cotton. This is a 3% decrease from last year's number, but 27% higher than the acreage of 2 years ago. Planting in Georgia started extremely slow due to a severely dry spring, but by June 1 was nearly on pace with average. Conversely, Texas experienced a near normal planting season although some replanting was necessary due to wind and hail damage .

CONCLUSION

Cotton nonwovens can be recycled, re-used or disposed off by natural degradation conditions. Cotton is a readily renewable resource with long-term supply assurance. Extensive research works is improving bleached fiber quality and quantity.

Nonwoven industries are producing various types of nonwovens with different manufacturing techniques, for better production. Cotton share of the textile fiber market has been steadily increasing and will continue to increase as cotton-containing items is preferred by the consumers.

PRODUCT DESCRIPTION

Cotton belongs to the category fibers/fibrous materials, which are classified as follows :

Plant hairs:

- Cotton seed-hairs
- Kapok tree fruit hairs

Stalk fibers from dicotyledonous plants (soft fibers):

- Flax, ramie (fine spinnable fibers)
- Hemp, jute, kenaf (coarse spinnable fibers)

Leaf fibers from monocotyledonous plants (hard fibers):

- Sisal, Manila hemp, palm fibers (poor spinning characteristics)

Bast:

- Linden, raffia palm, willow
- Basketwork material:
- Coconut fiber, rattan cane, halfa, piassava, esparto

Cotton consists of the unicellular seed-hairs of the bolls of the cotton plant (Gossypium hirsutum), which belong to the "plant hair" category. The cotton plant itself belongs to the mallow (Malvaceae) family. The fruits of the cotton plant burst when ripe, revealing a fist-sized tuft of cotton consisting of fibers up to 50 mm in length. Once picked, the cotton is dried in the sun and ginned (separation of seeds from fibers).

The plant fibers have a pronounced three-walled structure. The outer wax layer protects the primary wall. The most important element is the secondary wall, which consists predominantly of cellulose. The tertiary wall surrounds the lumen, which, in all cellulose materials of plant origin, is very well formed and filled with air.

The chemical composition of cotton is as follows:

- cellulose 91.00%
- water 7.85%
- protoplasm, pectins 0.55%
- waxes, fatty substances 0.40%
- mineral salts 0.20%

QUALITY / DURATION OF STORAGE

The quality of cotton is assessed according to the following criteria:

- Grade: cleanness, leaf content and other foreign matter of vegetable origin
- Color: color defects and staining
- Nature: ripeness, nep and nub content, uniformity
- Staple: composition of fiber material with regard to length
- Micronaire: fiber fineness
- Pressley: fiber strength

Color variations and loss of luster may indicate mold or bacterial attack. At 25 - 35°C and a relative humidity of 80 - 90%, color variations caused by microorganisms may be observed after as little as 3 - 4 weeks. Moisture-damaged bales must not be accepted.

Cotton is classified as follows according to staple size:

- Long-staple cotton: > 35 mm, e.g. Egyptian cotton, maco cotton; fine, with a silky luster and generally cream-colored.
- Medium-staple cotton: 25 - 35 mm, e.g. upland cotton (80% of world production), American cotton; white to yellowy-white.
- Short-staple cotton: < 25 mm, e.g. Indian cotton; generally yellowy-white to brownish and of lower quality, since often nonuniform and hard.

In addition to the long fibers, cotton also contains so-called cotton linters. These are short felt-like hairs (2 - 4 mm), obtained by further ginning of the seeds. Owing to their short staple length, they are not suitable for spinning, but instead serve as a starting material for cellulose-based manmade fibers and filaments and for cotton wool production.

The most valuable type is long-staple, fine-fiber cotton. Long cotton fibers have good spinning characteristics, strength, dyeability and resistance. Because it is so readily spinnable, cotton is the most important, most widely processed natural fiber of plant origin.

INTENDED USE

Cotton is primarily used in textile production, specifically for bed, household and table linen and for ready-made garments.

Table. Countries of origin

Europe	Bulgaria, Spain, Russia, Turkey
Africa	Egypt, Ivory Coast, Mali, Burkina Faso, South Africa, Sudan, Chad, Zimbabwe, Ethiopia, Algeria, Uganda
Asia	Burma, India, Pakistan, China, Iraq, Syria, Afghanistan, Iran
America	El Salvador, Guatemala, Mexico, Nicaragua, USA, Honduras, Argentina, Brazil, Venezuela
Australia	

PACKAGING

Cotton is usually shipped in largely square bales compressed to different degrees. The bales are tied firmly with steel straps or wire.

Strapping is essential to maintain compression of the bales during transport. If the strapping is damaged or broken, compression is diminished, which at the same time results in an increased supply of oxygen to the inside of the bales. This in turn increases the risk of combustion or feeds a fire which has already started. Bursting or chafing of the steel straps and wires may lead to sparking and external ignition. To protect the cotton bales from contamination and damp, they are wrapped in jute or plastic fabric, or sometimes in plastic films perforated so as to regulate bale moisture content. Bales vary in size and weight between 100 and 330 kg depending on country of origin.

CONTAINER TRANSPORT

Standard containers, subject to compliance with water content of goods, packaging and flooring.

CARGO HANDLING

In damp weather (rain, snow), the cargo must be protected from moisture, since cotton is strongly hygroscopic and readily absorbs moisture. This may lead to discoloration, decay, mold, mildew stains and rot. In addition, the cotton may swell by absorbing water vapor, resulting in an increase in volume of 40 - 45%. A high water content is difficult to detect from outside, since the cotton does not feel damp even with a water content of 20%.

Do not use hooks for cargo handling, since they may lead to sparking when they come into contact with the strapping.

In addition, smoking is absolutely prohibited during cargo handling.

STOWAGE FACTOR

- 1.42 - 4.60 m^3/t
- 1.39 - 3.76 m^3/t
- 1.70 - 4.25 m^3/t (bales)

STOWAGE SPACE REQUIREMENTS

Cool, dry

SEGREGATION

Fiber rope, thin fiber nets, jute strips

CARGO SECURING

The cargo is to be secured in such a way that the bales or strapping are not damaged. Undamaged strapping is essential to maintain compression of the bales during transport. If the strapping is broken, compression is diminished, which at the same time results in an increased supply of oxygen to the inside of the bales. This in turn increases the risk of combustion or feeds a fire which has already started. Bursting or chafing of the steel straps and wires may lead to sparking and external ignition.

RISK FACTORS AND LOSS PREVENTION

RF TEMPERATURE

Cotton requires particular temperature, humidity and possibly ventilation conditions (SC VI)(storage climate conditions).

Designation	Temperature range
Favorable travel temperature range	no lower limit - < 25°C
Optimum travel temperature	20°C
Autoignition temperature (for oily cotton)	120°C
Glow temperature	205°C
Fire point	210°C
Ignition temperature	407°C

At temperatures > 25°C, cotton dries out, becomes hard and brittle and losses elasticity. Light causes the same deterioration. The optimum temperature for mold development is 25 - 35°C.

Cotton is subject to self-heating/spontaneous combustion. The autoignition temperature of oily cotton is 120°C.

At temperatures < 0°C there is no risk of wet bales rotting, since this process stops at low temperatures. In some cases, damaged cotton has been placed in intermediate cold storage, so preventing rot.

Every hold should be equipped with means for measuring temperature. Measurements must be performed and recorded daily.

RF HUMIDITY/MOISTURE

Cotton requires particular temperature, humidity and possibly ventilation conditions (SC VI)(storage climate conditions).

Designation	Humidity/water content
Relative humidity	65%
Water content	7.85 - 8.5%
	8.5%
Maximum equilibrium moisture content	65%

Cotton behaves strongly hygroscopically (hygroscopicity). It must be protected from sea, rain and condensation water and also from high levels of relative humidity, if decay, discoloration, mold, mildew stains and rot are to be avoided.

Damage to cotton due to moisture is divided into "country damage" and "heart damage. "Country damage" may arise in the country of origin as a result of water penetrating from outside or soiling with dirt, mud or sand. Such damage may be detected from outside and may reach right into the bales. "Heart damage" arises as a result of compression performed under excessively moist conditions during wrapping and is not visible from outside. The cotton bales may swell by absorbing water vapor, resulting in an increase in volume of approx. 40 - 45%. At a relative humidity of 95%, cotton may increase its water content to 25 - 27% without feeling wet. For this reason, moisture levels should be measured when a consignment is accepted. Moisture-damaged bales must not be accepted.

RF VENTILATION

Cotton requires particular temperature, humidity and possibly ventilation conditions (SC VI)(storage climate conditions).

If the product is loaded for shipment in a dry state, it does not have any particular ventilation requirements. Problems arise if the product, packaging and/or flooring/ceiling are too damp or the temperature is too high. The heat has then to be dissipated and ventilation should proceed as follows:

Air exchange rate: 6 changes/hour (airing)

Moisture must then be eliminated, to reduce mold and bacterial activity.

Since cotton very readily absorbs oxygen, before anybody enters the hold, it must be ventilated and a gas measurement carried out, since a shortage of oxygen may endanger life.

RF BIOTIC ACTIVITY

Cotton displays 3rd order biotic activity. It belongs to the class of goods in which respiration processes are suspended, but in which biochemical, microbial and other decomposition processes still proceed.

RF GASES

Cotton very readily absorbs oxygen. An oxygen shortage may therefore arise in closed holds. Before anybody enters such holds, the holds must be ventilated and, if necessary, a gas measurement carried out.

The increase in CO_2 and CO content indicates a cargo fire. The TLV of the hold air is 0.49 vol.%. As a result of the oxygen-rich lumen (cavity in fiber), bales often burn for weeks without being discovered. The high absorption capacity of the fibers means that the gaseous combustion products are soaked up by the cotton fibers and the smoke alarm is often triggered too late.

RF SELF-HEATING / SPONTANEOUS COMBUSTION

Cotton has an oil content of 0.4 - 1.0% (waxes).

Cotton is assigned to Class 4.1 of the IMDG Code (Flammable solids).

However, its specific characteristics and negative external influences. Its high cellulose content makes cotton particularly liable to catch fire through external ignition.

Therefore, it must always be protected from sparks, fire, naked lights and lit cigarettes. Smoking is absolutely prohibited. Sparks may arise from bursting or chafing of the steel straps (and also as a result of inadequate cargo securing in the hold or container) and cause a cargo fire. In accordance with the IMDG Code, ventilation openings leading into the hold should be provided with spark-proof wire cloth. In the case of fire caused by adjacent cargo, the hydraulically compressed bales burn for a long period. The considerable compression prevents the fire from spreading as quickly as it would spread through uncompressed bales. External fire damage is generally limited to a superficial 1.5 cm deep carbonized layer. Where saw gins are used, external ignition by sparks may be caused even before the cargo is accepted. During such ginning, friction-induced sparks may reach the inside of the bale, causing burning over a relatively long period.

In addition to external ignition, cotton is also liable to thermal, chemical and microbial self-heating/spontaneous combustion. Self-heating/spontaneous combustion arises as a result of moisture, fats/oils, due to the action of acids, such as nitric or sulfuric acid, and through contact with oxidizing agents and with goods with a tendency to self-heating. As a result of the very well developed oxygen-rich lumen (cavity in the fiber) of the cotton fiber and the oxygen supply contained in the capillary cavity system, smoldering fires inside the bales often last for weeks. Self-heating is also caused by microbes, which may produce small amounts of methane gas if the bales were baled when wet.

Traces of vegetable oil, such as the wax content of the fiber, or quick-drying oils (castor oil, linseed oil, tung oil, hemp-seed oil, poppy-seed oil etc) cause self-heating/spontaneous combustion, without the need for any oxygen from external sources. Therefore, do not stow cotton with fats/oil, or with oil-bearing seeds/fruits. Vegetable oils are easily oxidized on cotton, producing a risk of spontaneous combustion. The oil absorbed by the fiber and its subsequent vaporization may easily result in an explosive mixture of air and oil vapor in unventilated corners of the hold or container.

Fire-fighting is best performed using CO_2. It is very difficult to extinguish a fire because of the excess of oxygen in the cotton fiber, which maintains the fire from the inside. When fighting a fire, do not break the steel straps or open the bales, since relieving the compression increases the oxygen supply and makes it impossible to fight the fire effectively.

Water must not be used for fire-fighting, since the swelling capacity of the cotton fibers (40 - 45% increase in volume) may cause damage to the hold or container walls. Elevated temperatures increase the level of swelling, thereby increasing rigidity, which can have very inconvenient consequences if water is used to extinguish a fire or gets in otherwise: a batch of cargo may swell until jammed, so that the bales have subsequently to be wrenched out.

RF ODOR

Active behavior	Cotton has a slight, pleasant odor. A conspicuous musty odor indicates mold and rot processes inside the bales.
Passive behavior	Cotton is sensitive to unpleasant or pungent odors. Odor taint may be caused by, among other things, kerosene, diesel fuels, fish oil, fish meal, stabilized meal and naphthalene.

RF CONTAMINATION

Active behavior	Cotton does not cause contamination.
Passive behavior	Cotton is sensitive to contamination by dust, dirt, fats/oils and rust as well as oil-containing goods, such as oil-bearing seeds/fruits, copra, raw wool etc., since oil-impregnated fibers promote self-heating/cargo fire. Increased contamination of cotton provides microorganisms with an excellent nutrient medium, which means that holds or containers must be suitably clean and in a thoroughly hygienic condition. Residues from previous cargoes, such as ores, stones, coal, metal filings, fertilizers etc., result in losses. Rust contamination may be caused by rusty steel straps, among other things. Since rust hampers the spinning process, this represents a reduction in value. Reddish-brown discoloration of the bales indicates "country damage" by red soil dust (tropical weathered soil, laterite) which penetrates through the wrapping. Further contamination may occur: **Contamination by dust/sand:** arises in conjunction with mostly dried damp on sides, tops/bottoms not covered with wrapping. Affected cotton is dirty to discolored, brittle and musty. This damage may occur at any time prior to shipment, up to the time of loading. **Contamination by red-brown soil dust:** ore dust (sometimes containing minerals) may lead to considerable discoloration of cotton on sides, tops/bottoms not covered by wrappings or sometimes by penetrating through the wrappings. This damage is caused during transport/storage prior to shipment. **Contamination by grass/straw:** This occurs on sides not covered by wrappings and is caused during storage prior to shipment. This contamination may be removed by brushing. **Color contamination:** contamination with colors of all kinds may be caused prior to shipment and on board. In addition, damage has been noted which is caused by excessively moist application of marking, which is sometimes undertaken by consignors but may also be carried out on board for the purpose of segregating stowed loads.

RF MECHANICAL INFLUENCES

Care must be taken to ensure that mechanical influences do not cause damage to strapping, which increases the risk of fire by relieving the compression of the bale and allowing a greater supply of oxygen. Use no hooks.

RF TOXICITY / HAZARDS TO HEALTH

Since cotton is highly oxygen-absorbent, a life-threatening shortage of oxygen may arise in the hold. Thus, before anybody enters the hold, it must be ventilated and, if necessary, a gas measurement carried out. The TLV for CO_2 concentration is 0.49 vol.%.

RF SHRINKAGE/SHORTAGE

Loss of volume may be caused by tally errors.

RF INSECT INFESTATION / DISEASES

Insects, in particular ants and beetles, may damage the bales during storage ashore. So-called honeydew is deposited as a secretion on the cotton by an insect (white fly). This secretion contains sugar and makes the cotton fibers sticky. Honeydew is barely visible to the eye.

Mold growth caused by heat and moisture may start even in the cotton field. This leads to a reduction in value by staining and discoloration due to rot.

Cotton exhibits low resistance to bacterial degradation and mold growth. Mycelial fungi cause circular mildew stains which are gray/yellow/green, orange/red and brown/black in color, together with a musty odor. Particularly active on cotton is the mold Stachybotrys sporium, which penetrates into the lumen of the cotton fiber. Within 10 days, the cotton loses 30% of its strength by cellulose degradation. Mold and bacteria change the color of the cotton to yellow, yellow/green, orange, red to chestnut-brown and gray. This is accompanied by a loss of luster. At 25 - 35°C and a relative humidity of 80 - 90%, these variations may be observed after 3 - 4 weeks.

According to , approximately 50% of world cotton production is destroyed by parasites and diseases.

MONITORING NITROGEN NUTRITION IN COTTON

Appropriate N rates increase productivity and improve fiber length and strength and are essential for cotton growth and flowering, but N excess may induce rank-growth, extend the plant cycle and decrease lint yield and fiber quality (Staut & Kurihara, 2001). The intensity of N uptake is low in the first 35-40 days, until the appearance of the first flower buds (Rosolem, 2001). Thereafter, N uptake increases to around 5.5 kg ha^{-1} day^{-1}75-80 days after plant emergence (DAE), and then N uptake declines with plant age. However, Rosolem & Mikkelsen (1989) concluded that while a significant number of bolls are developing, there may be potential for response to N, but the nutrient taken up after 90 DAE accumulates mainly in the leaves of the median and upper plant parts. In well-nourished plants, less than 30 % of N absorbed at this time will be found in the fruits.

Nitrogen is the nutrient most required by cotton, but the empirically recommended N fertilization in Brazil has not been based on plant response to N or soil tests. Since both N excess and deficiency can lead to losses in cotton yield and quality, regular and accurate determinations of the nutritional status underlying the N management fertilization would be useful, as commonly used in cotton-growing areas around the world. Foliar diagnosis is important in monitoring the nutritional status of plants, but the analysis is time-consuming.

The concentration of NO_3^- in cotton petioles decreases from the first week before flowering until the third week after the beginning of flowering (Mozaffari et al., 2004), but during the flowering period it is correlated with cotton yield. By regular petiole samplings during this period it is possible to detect and correct N deficiencies. One problem is that the amount of water available and the position of the node on the main stem affect the amount of N detected in the petiole from the third week after the appearance of the first floral bud until the fourth week after the first flower opens. Measurements of petiole NO_3^- one week before or after early flowering is a sensitive indicator of applied N rate and may be a good predictor of plant growth and lint yield within a season.

The portable ion-specific NO_3^- meter Cardy® (C-141 Cardy Nitrate Meter, Horiba, Inc.) allows direct and fast readings from a few drops of the material (sap or soil solution). Hochmuth et al. (2003) used Cardy® measurements on petiole extracts of basil plants, showing the reliability of the results in predicting plant N status. The concentrations in soil solution and the sap of tomato plants obtained with an ion-specific nitrate meter for NO_3^-, K^+ and Na^+ were well-correlated with results obtained through traditional measurements.

Indirect estimation methods of the plant N status, e.g., the chlorophyll content, have been evaluated. The Minolta chlorophyll meter SPAD-502® (Minolta, 1989) provides readings (SPAD units) which, within limits, are directly correlated with chlorophyll content and tissue N concentration. According to Sibley et al. (1994) the values determined with SPAD-502 do not depend on or are not affected by the environmental brightness levels. The results of the application of the indirect chlorophyll meter SPAD-502 have been satisfactory for the assessment of the N nutritional status of some crops. Hence, the evaluation of N nutritional status of cotton could be done with the SPAD-502, using readings made in leaves of the middle third and top of the plants. Malavolta et al. (2004) found a positive correlation between chlorophyll concentration (SPAD readings) and cotton lint yields. According to Malavolta (2006), cotton leaf samples for chlorophyll content analysis with the SPAD meter must be collected at full bloom, and the meter must be placed on the side lobes of the 4th or 5^{th}leaf from the apex. The purpose of this study was to investigate the correlation between two methods used for rapid determination of N nutritional status in cotton, a chlorophyll meter (Minolta SPAD-502®) and an ion-specific NO_3^- meter (C-141 Cardy® Nitrate Meter, Horiba, Inc.) with N contents determined in laboratory (conventional analysis of dry tissue), as well as their correlation with cotton yield.

MATERIALS AND METHODS

An experiment was conducted under field conditions on the farm Santa Tereza in Itapeva, State of São Paulo, Brazil, in a Clay Red Latossol (Oliveira et al., 1999). Soil analysis showed pH 5.8, 22 g $dm^{-3}$of OM, 17 mg dm^{-3} of P_{resin}, 3.1, 21 and 13 $mmol_c$ dm^{-3} of K, Ca and Mg, respectively, and 58 % of base saturation. Rains were monitored and observed data.

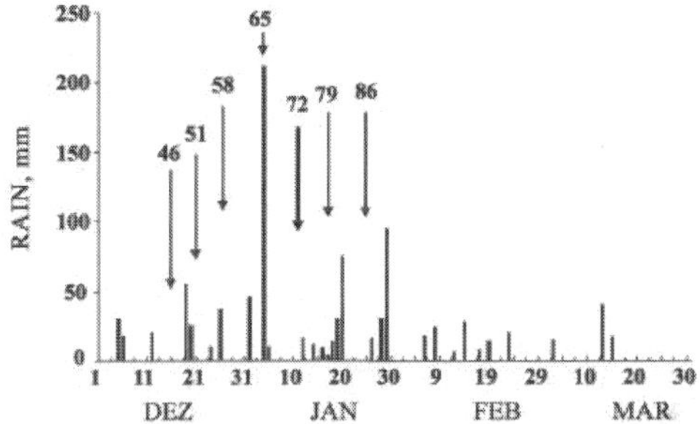

Figure 1. Rain at the experimental site during cotton growth. Arrows show sampling times and the labels are days after plant emergence in each sampling.

Cotton (*Gossypium hirsutum* L., var. Latifolia) cv. Delta 90 BT was planted on 28 October 2007, using 500 kg ha^{-1} of the fertilizer 8:28:12 applied to the seed furrow. Stabilization of the number of plants emerging was observed five days after planting, which was considered the date of plant emergence. Four DAE, 80 kg ha^{-1} K_2O was applied, along with part of the N, according to the treatments. The crop management normally used by the farmer was applied, with the exception of N application. Cotton was harvested by hand on April 28, 2008.

Table 1. Rates and time of N application to cotton

	N rates				
	kg ha^{-1}				
Planting	40	40	40	40	40
4 days after plant emergence	0	15	30	45	60
58 days after plant emergence	0	15	30	45	60
Total	40	70	100	130	160

Treatments consisted of the following doses of sidedressed N: 0, 30, 60, 90, and 120 kg ha^{-1}. The first application was made as close to planting as possible, in order to establish different levels of N nutrition from the beginning of the plant cycle. The following methods to assess the nutritional status of cotton in N were studied: a chlorophyll meter (Minolta SPAD-502®); an ion-specific NO_3^- meter (Nitrate Meter C-141 Horiba-Cardy®) and conventional analysis of dry tissue in the laboratory.

The assessment of the plant N nutritional status with the chlorophyll meter is not destructive. Chlorophyll readings of eight plants per plot were taken from the fourth and fifth uppermost fully expanded leaves on the main stem of cotton. The plants were sampled weekly for seven weeks, beginning one week after the onset of "pinhead". The petioles of the same leaves were collected for ion-selective NO_3^- readings. The petioles were cut into pieces approximately 2 mm, which were placed in a 3 mL syringe. The sap was extracted by applying pressure on the plunger and four to five drops were placed directly on the sensor pad of a Cardy nitrate meter for analyses. For laboratory analyses the forth and fifth uppermost leaves on the main stem of cotton were sampled from 10 plants per plot, dried in a forced air oven at 65 °C for three days, ground and N determined by Kjeldahl steam distillation.

The experimental design was a randomized complete block with four replications. The plots consisted of six 6-m rows of cotton, separated by a distance of 1.0 m. A correlation study was run on the weekly results obtained with the chlorophyll meter, the selective-ion meter and lab determinations. The standard error was calculated for each determination method in each week. Regressions were adjusted between cotton production and chlorophyll readings, NO_3^- and N concentrations.

RESULTS AND DISCUSSION

Cotton yields ranged from 3,500 kg ha^{-1} to more than 5,000 kg ha^{-1}, compared with the state average of 2,900 kg ha^{-1}. As expected, there was a significant linear response to side dressed N rates up to 120 kg ha^{-1}, corresponding to a total application of 160 kg ha^{-1} N.

The NO_3^- levels determined with the selective-ion meter correlated with SPAD readings from 58 DAE onwards, or from the second week of flowering, and with N contents in leaves 58 and 65 DAE (second and third week of flowering). However, the NO_3^- content was correlated with cotton yields before the third week of flowering (46 DAE). According to Smith et al. (2000), the NO_3^--N concentrations in the petiole sap were highly correlated with dry petiole NO_3-N contents throughout the season. Sunderman et al. (1979) stated that petiole NO_3^- content, one week before or after early flowering, is a sensitive indicator of applied N rate and may be a good predictor of plant growth and lint yield of a growing season. Our results show that the correlation with cotton yields may be high from one week after pinhead to the fifth week of flowering.

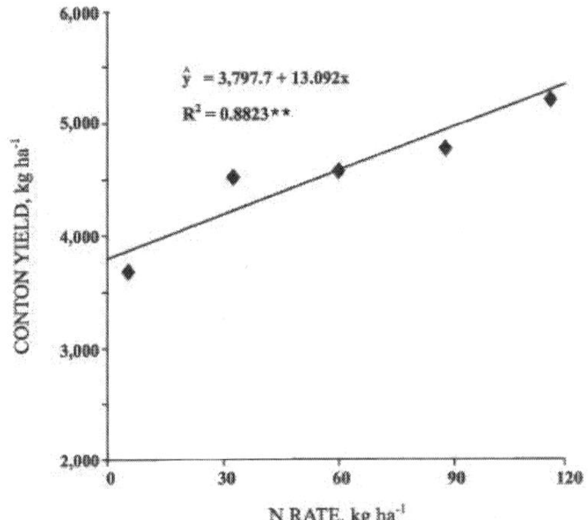

Figure 2. Cotton yields as affected by sidedressed nitrogen rates. Santa Tereza Farm, 2008. Each dot represents the average over four replications. *, R^2 significant ($p < 0.05$)

Chlorophyll contents were significantly correlated with cotton production from the third to the fifth week of flowering. SPAD readings should be directly correlated with chlorophyll content and tissue N concentration, in a certain range.

However, this was not observed in our experiment, where the correlation between chlorophyll contents and N concentration was only significant in the third week of flowering.

Nitrogen concentrations in leaves determined in the laboratory were correlated with cotton yields, except in sampling one week after pinhead (46 DAE) and full flowering (79 DAE). Before the appearance of the first flower a correlation was not expected.

But it is rather difficult to explain the lack of correlation during full bloom, the period in which sampling for foliar diagnosis is recommended (Rosolem et al., 2007). One possible reason for the lack of correlation is the heavy rain observed in this period.

In general, N contents determined in cotton leaves increased with N rates, as expected. It is also interesting that N concentrations in the diagnostic leaf increased from the second to the sixth sampling, decreasing afterwards. This period begins with the appearance of the first flowers and lasts until full bloom, usually when sampling for foliar diagnosis is recommended. For maximum productivity, N leaf contents at cotton full bloom should range from 40 to 45 g kg^{-1}. In this study N concentration in the diagnostic leaf were within or above the threshold level and a response to N rates was observed even when sampled earlier than recommended for foliar diagnosis.

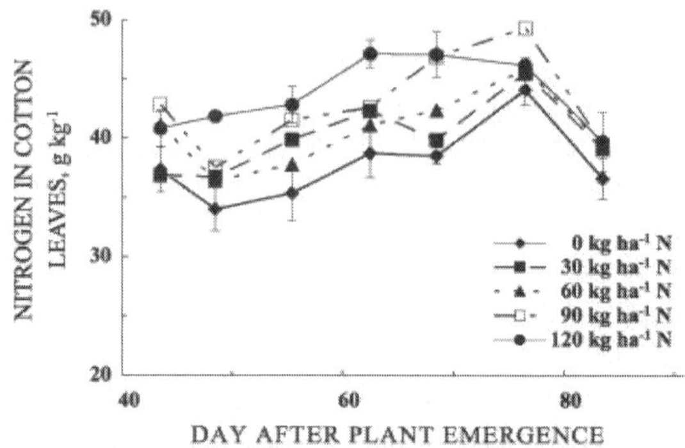

Figure 3. Nitrogen concentrations in the fourth and fifth uppermost leaves on the main stem of cotton, determined in laboratory, as affected by N rates and sampling time. Vertical bars show the mean standard deviation (bars are shown for rates of 0 and 120 kg ha^{-1} only, for clarity). Average of four replications.

Leaf N contents were significantly correlated with cotton yields from one week before flowering to the second week of flowering, and again in the fifth week of flowering, or full bloom, when sampling for foliar diagnosis is recommended (Malavolta, 2006). In the third and fourth weeks of flowering, 72 and 79 DAE, the correlation was not significant. One possible explanation for the lack of correlation in these samplings would be the effect of rain, but in this case a fertigation was applied at 72 DAE and at 79 DAE the soil was water logged, which may have, which may have interfered with the results. Nitrogen levels in the fourth and fifth uppermost leaves on the main stem could therefore be used to diagnose cotton N nutrition as early as one week before the appearance of the first flowers, which could be important for N management since a timely correction of the fertilization program would be possible.

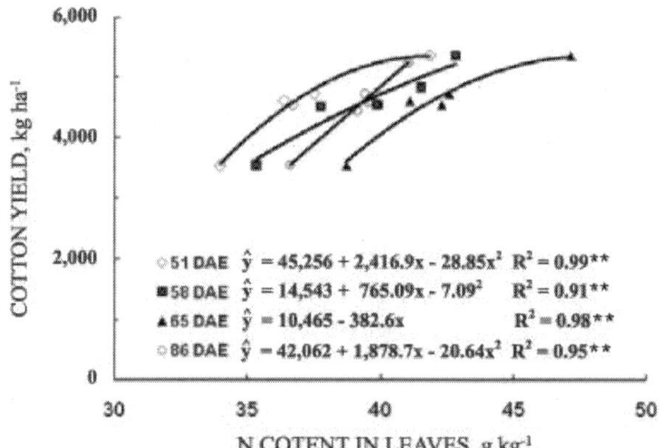

Figure 4. Regressions between cotton yields and N contents in cotton leaves, as affected by sampling time. **, R^2 significant ($p < 0.01$).

The NO_3^- contents in cotton petioles, determined by the selective-ion meter, responded to the N rates, although the correlations of N fertilizer rates and cotton yields were not always significant. The NO_3^- levels in cotton petioles ranged from less than 2,000 to more than 9,500 mg L^{-1}. Smith et al. (2000) found that NO_3^- determined in dry petiole was highly correlated with NO_3^- in petiole sap and the equation describing this relationship is = -1,170.86 + 9.96x, where y is NO_3^- in dry petiole (mg kg^{-1}) and x is NO_3^- in petiole sap (mg L^{-1}). Therefore, considering a 10-fold relationship between petiole sap and dry petiole NO_3^-, NO_3^- in dry petiole would range approximately from 200 to 950 mg kg^{-1} in our experiment. Petiole N-NO_3^- concentrations as high as 24-30 g kg^{-1} have been reported in upland cotton, but a critical minimum level of 1-2 g kg^{-1} during the first flower period was suggested to ensure growth and yield of upland cotton in California. In Texas, this level was adequate for maximum production of Pima cotton, depending on the season.

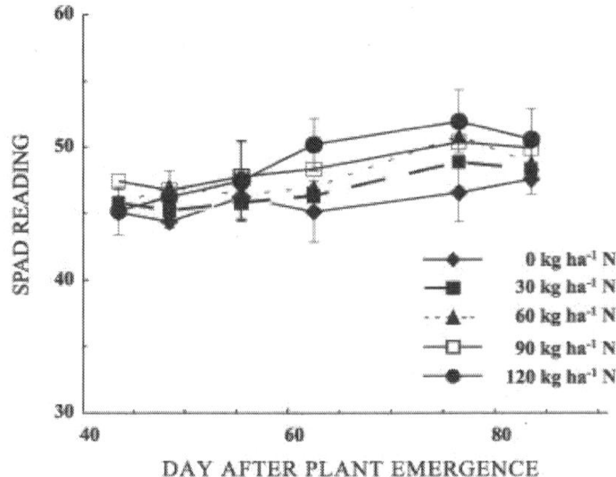

Figure 5. NO_3^- levels in cotton fourth and fifth uppermost main stem petioles, determined with the specific ion meter CARDY-NO_3^-, as affected by N rates and sampling time. Vertical bars show the mean standard deviation (bars are shown just for rates of 0 and 120 kg ha^{-1} for clarity). Average of four replications.

In Arkansa, USA, it was observed that NO_3^- concentration in cotton petioles decreased from the first week before flowering to full bloom, during the fourth or fifth week of flowering (Mozaffari et al., 2004). In the present experiment petiole NO_3^- decreased from one week from pinhead to the fourth week of flowering in malnourished plants. For well-nourished plants, petiole NO_3^- initially increased until the first week of flowering and then decreased for three weeks. Regressions between petiole NO_3^- concentrations as determined with the selective-ion meter and cotton yields were adjusted for from two weeks before the appearance of flower buds until the second week of flowering. Malavolta et al. (2004) found high correlation between petiole NO_3^- levels and cotton production 54 DAE (just before flowering), which supports our results. Levels of N-NO_3^--N in the cotton petiole in the order of 8000 - 10,000 mg L^{-1} were associated with the highest yield in this study, under field conditions, whereas in nutrient solution, the level corresponding to the highest yield was 12,300 mg L^{-1} (Malavolta et al., 2004), which would correspond to approximately 1.2 g kg^{-1}(Smith et al., 2000).

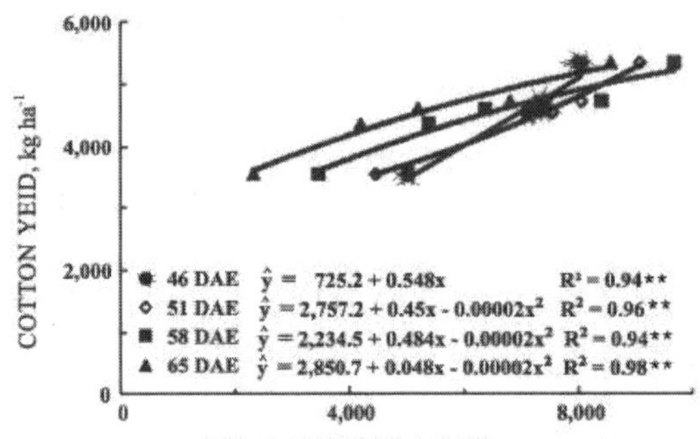

Figure 6. Regressions between cotton yields and NO_3^- contends in cotton petioles, as affected by sampling time. **. R^2 significant ($p < 0.01$).

As to the use of chlorophyll readings to estimate N status, there was a significant response to N rates from the second week of flowering onwards. As it was observed for N tissue concentrations this method was also sensitive to changes in N availability before the full bloom stage, when sampling for foliar diagnosis is recommended. In a nutrient solution experiment, Malavolta et al. (2004) observed a significant correlation between SPAD index readings and levels of available N from the second week of flowering, which is similar to the results of this study.

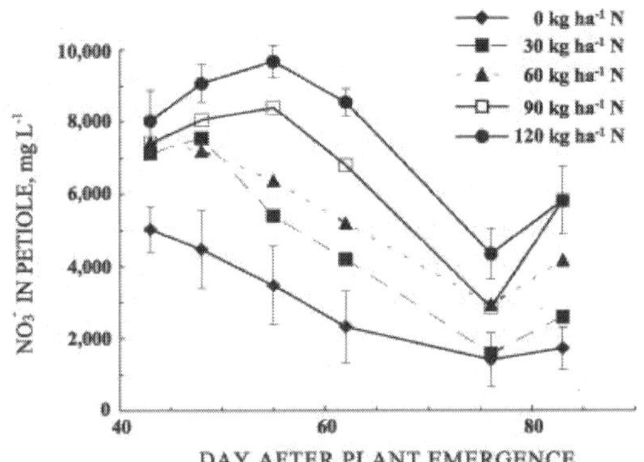

Figure 7. SPAD index readings in the fourth and fiftl uppermost leaves on the main stem of cotton as affected by N rates and sampling time Vertical bars show the mean standard deviatioı (bars are only shown for rates of 0 and 120 kg ha⁻ for clarity). Average of four replications.

In nutrient solution, where the N availability levels differed greatly, Malavolta et al. (2004) concluded that SPAD readings during the flowering period were effective in determining the nutritional status of the plant and were related with cotton yields.

In their experiment, maximum production was observed when readings were around 50, during flowering. In the present experiment, SPAD readings were similar at N rates from 100 kg ha^{-1} and from the fourth sampling onwards, which is coincident with results of Malavolta et al. (2004). In our experiment, the yield was highest when the SPAD index was greater than 50. Motomiya et al. (2007) also observed a good correlation between N concentrations in cotton leaves and SPAD readings, reaching maximum levels (> 35 g kg^{-1}) at readings above 50.

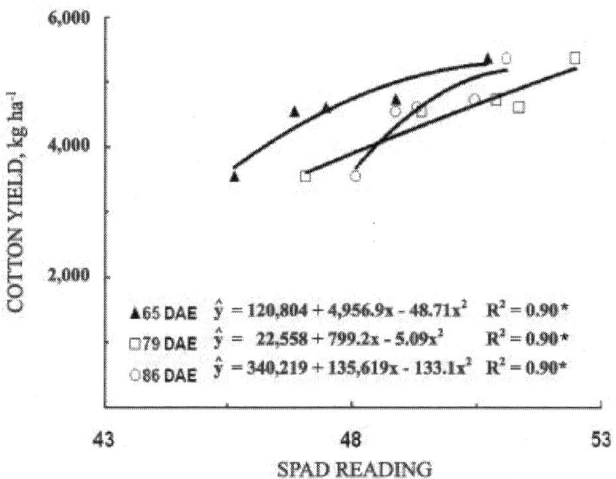

Figure 8. Regressions between cotton yields and SPAD readings in cotton leaves, as affected by sampling time. *, R^2 significant ($p < 0.05$).

4

Cotton Production and Marketing

The world cotton industry has experienced dramatic changes over the last five decades as production nearly quadrupled, rising from 6.6 million tons in 1950/51 to a record of 26.3 million tons in 2004/05. The average rate of growth in world production over the last five decades has been about 2.5% per year, or about 280,000 tons per year. Growth in cotton production was steady during the 1950s and 1960s but slowed during the 1970s because of slower world economic growth and limited gains in cotton yields. World cotton production exploded from 14 million tons in the early 1980s to 19 million tons in 1984/85, as market incentives and the widespread use of better seed varieties and better methods of plant protection led to increased yields. World production climbed to a record of nearly 21 million tons in 1991/92 but levelled off during the 1990s. With the commercial application of biotech cotton varieties beginning in 1996 and the expansion of cotton areas in francophone Africa, Australia, central Brazil, western China, and Turkey, world production exceeded 26 million tons in 2004/05 and has remained nearly as high during the two seasons since.

Since 1950/51, the world area dedicated to cotton has fluctuated between 28 million hectares and 36 million hectares; the average has been 32.7 million hectares. While there have been dramatic reductions in cotton area in some regions since the 1950s, particularly in the United States, North Brazil and North Africa, there have been offsetting increases in francophone Africa, Australia, China, India, Pakistan and the Middle East. With total area showing no tendency to rise, all the growth in world cotton production since the 1940s has come from improved yields.

The world cotton yield in the early 1950s was 230 kilograms of lint per hectare. Yields rose steadily at an average rate of more than 2% per year during the 1950s and 1960s, and then grew more slowly from the mid-1970s until the mid-1980s. During the 1980s the world cotton yield rose dramatically, reaching a record of nearly 600 kilograms per hectare in 1991/92. However, yields stagnated during the 1990s due to problems associated with diseases, resistance to pesticides, and disruption of production for economic reasons. Yields began rising again in the late 1990s with improvements in seed varieties and the use of biotech varieties, and the world yield in 2004/05 reached a record 747 kilograms per hectare. The yield in 2006/07 is estimated at essentially the same level, and still well above the previous five-year average. The average rate of increase over the last six decades has been a little more than 8 kilograms per hectare per year.

PRODUCTION TECHNOLOGY OF COTTON

Climate- Cotton is a crop of subtropical climate. Cotton needs on an average a minimum temperature of 60^0 F for germination, 70 -80 ^{0}F for vegetative growth, 80-90^{0}F with cool nights during fruiting period.

An annual rainfall of at least 50 cm is minimum requirement for cotton unless it is grown on irrigated soils. Ultimately rains and the heavy humid weather during later stages of cotton season may spoil the produce, lower its ginning properties or promote attack of insect, pest, diseases. So weather should be clear at harvesting because rain at this stage will discolour the lint and reduce its quality.

Soil- Cotton needs a soil with a excellent water holding capacity and aeration and good drainage as it cannot withstand excessive moisture and water logging. The major group of soil for cotton cultivation are the alluvial soils, black soils, red sand loam

Seed rate and spacing- Depending upon the variety, soil type, the cultivation on practices and method of sowing, seed rates and spacing have been recommended.

The seed rate of 15-25 kg/ha and spacing of 75-90 cm between the rows are generally recommended for irrigated conditions. For rainfed deshi-cotton seedrate of 12-16 kg/ha and spacing 45-60 cm between rows are adopted. For rainfed American cotton seedrate is 12-16 kg and spacing is 60-75 cm between the rows.

Optimum sowing time- Sowing of crops depends upon water resources. If irrigation facilities are available the crop should be sown in April, especially American cottons, otherwise it must be sown just after the monsoon starts but not later than 15th July. Sowing in rows can be done either by drilling, dibbling, or placing the seeds in furrows behind the country ploughs.

Manures- Fern yard manures rarely applied to cotton in the states of Punjab, Haryana, Rajasthan, Madhya Pradesh. In Maharashtra the cotton crop is manured with farm yard manures once in 3-4 years @ 12-15 tonnes/ha.

Fertilizers- For rainfed crop 20 kg of N 18 kg of P and 78 kg K is economical Nitrogen is applied in split doses, half dose at the time of sowing and other half as top dressing thinning or just before flowering.

For irrigated cotton the dose is double.

Water requirement- In northern India the irrigated cotton crop is mostly sown after a preliminary heavy irrigation and second light irrigation is given three to four weeks after germination. Subsequent watering depends upon the nature of the soil and the weather conditions.

Flowering and ball formation are the critical stages from the point of irrigation. Inadequate irrigation stages during this stages leads to a heavy shedding of flower buds and bolls.

In Maharashtra cotton is sown on ridges and given two light irrigations immediately. Then after one or two irrigations are given and the crop is sustained till the onset of monsoon. Depending upon the rains one or two irrigations are necessary.

Generally crop needs 6-8 irrigations and 600-800 mm of water during its lifetime.

Cultivation practices:- Before sowing the soil is first given a heavy irrigation followed by one or two ploughings then the soil is planted with wooden plank In Central India the rainfed cotton, field is harrowed for 3-4 times.

Interculture- Weeding is followed after 30-40 days of sowing subsequently hoeings are given to control weeds. Thinning of cotton is a special feature of the irrigated crop.

Diseases and pests- Cotton aphids cotton jassides are controlled by spraying Malathion 0.08%.

Cotton leaf roller, spotted boll corn, pink boll corn are controlled by dusting crop with 10% carbonyl where red cotton bug and dusky cotton bug are controlled by dusting 5% B.H.C.

Optimum harvesting time- Cotton is harvested in three or more pickings taken suitable intervals. The season of harvesting varies with of sowing, duration of variety. Generally the crop is sown in June-September and is September-October is harvested from November to March to June respectively. Well dried bolls are picked.

Varietal Improvement- New improved cotton varieties were released for cotton cultivation in various tracts in India. Out of them, 320-F in Punjab, H-14 in Haryana, Deviraj and Digvijay in Gujrat and MCU-2 in Tamilnadu are noteworthy. In second plan Badnawar-1 in Madhya Pradesh, Buri-14 in Maharashtra, MCU-3 in Tamilnadu are some newly released strains.

PRODUCTION TRENDS

Cotton is produced in about a hundred countries, but production has traditionally concentrated in a few. Over the last three decades, the four leading producing countries have accounted for an increasing share of world production. China, the United States, India and Pakistan accounted for 48% of world production in 1970/71 and 72% in 2006/07. The share of industrialized countries (the United States, Australia, Spain and Greece) was little changed at 19% of world production in 1980/81 and 21% in 2006/07. Developing countries accounted for 61% of world production in 1980/81 and 72% in 2006/07. Cotton production in the former Soviet Union declined during the last two decades, accounting for 19% of world production in 1980/81 and 7% in 2006/07.

Production in China, the largest producer, increased at an average annual rate of 5% during the 1980s and fluctuated within a range of 3.7 to 5.7 million tons during the 1990s. Production in China rose to a record of 6.3 million tons in 2004/05 and is forecast to reach a new record of 6.7 million tons in 2006/07.

In the United States, cotton production increased from 2.4 million tons in 1980/81 to 3.3 million tons in 1990/91, and fluctuated between 3 and 4.3 million tons during the 1990s before rising to 5.2 million in 2005/06. United States production was 4.7 million tons in 2006/07 because of reduced rain.

Cotton production in India rose from 1.3 million tons in 1980/81 to 3.0 million tons in 1996/97. Thereafter, production fell to 2.3 million tons in 2002/03 before reaching a record of 4.6 million tons in 2006/07. India has rapidly adopted biotech cotton varieties since 2002, and this seems to be contributing to rising yields.

Production in Pakistan expanded rapidly during the 1980s, growing from 700,000 tons in 1980/81 to 2.2 million tons in 1991/92. However, production fell in 1992/93 and remained below the 1991/92 level until 2004/05 when production rose to 2.5 million tons. Difficulties combating disease are again resulting in lower production, forecast at 2.1 million tons in 2006/07.

Cotton production in Brazil declined rapidly between the mid-1980s and the mid-1990s because of structural changes in favour of soybeans and because of infestation from the cotton boll weevil. Brazilian production recovered in the second half of the 1990s as production shifted to the centre of the country, where cotton had not previously been grown. Production, which declined from 965,000 tons in 1984/85 to 310,000 tons in 1996/97, climbed back to 940,000 tons in 2000/01 and to 1.3 million tons in 2004/05, surpassing Uzbekistan and Turkey. Production in 2006/07 is estimated at 1.4 million tons.

Cotton production in Turkey increased from 650,000 tons in 1990/91 to 850,000 tons in 2006/07. Cotton production in Australia increased very rapidly during the 1980s and 1990s, from 100,000 tons in 1980/81 to 800,000 tons in 2000/01. Because of drought, production was only 250,000 tons in 2006/07. Cotton production in the European Union (EU) increased from 300,000 tons in 1990/91 to 500,000 tons in 2004/05. A combination of weather and changes in cotton policies in the EU led to a reduction in output to an estimated 370,000 tons in 2006/07.

AFRICA IN PERSPECTIVE

In Africa, cotton production increased from 1.3 million tons in 1990/91 to 1.8 million tons in 1997/98, but low cotton prices discouraged additional increases in African production in the following years. African production rose to 2 million tons in 2004/05 but fell to 1.6 million tons in 2006/07. Francophone countries in West and Central Africa produced 870,000 tons in 2006/07, accounting for 54% of production in the continent.

Cotton production in Africa as a whole rose by 3% per year from 1994/95 to 2005/06, after having been flat during the 1980s and early 1990s. The growth in African production coincided with a devaluation of the CFA franc. However, production has not grown in all countries. Production in North Africa has changed little since 1994/95, at 380,000 tons, while production in francophone Africa rose from 600,000 tons to 900,000 tons. Production in Southern and East Africa rose from 280,000 tons in 1994/95 to 460,000 tons in 2005/06.

However the expansion in African production came mostly from increases in area devoted to cotton, rather than rising yields. During the first three seasons of the 1980s, the average cotton yield across Africa was 336 kilograms per hectare, which equalled 78% of the world yield at that time of 433 kilograms per hectare. By the early 1990s, the average African yield had risen to 362 kilograms, but this was just 63% of the world yield. And, during the three most recent seasons, the average African cotton yield was barely changed at 369 kilograms per hectare, just 52% of the world yield of 705 kilograms. There are a number of reasons why yields in Africa have not risen in tandem with the world yield, including lack of access to inputs, weak research and extension systems in many countries, and the fact that very little irrigation is used on cotton in Africa, while more than half of world production is irrigated.

African cotton area rose from a three-year average of 3.5 million hectares in the early 1980s, representing 10% of world area, to an average of 4.9 million hectares currently, accounting for 14% of world area. Cotton area in East and Southern Africa fell from 1.9 million hectares in the early 1980s to 1.1 million hectares during one year in 1993/94, before rising again to 2 million hectares currently. The area devoted to cotton in North Africa fell from 880,000 hectares during the early 1980s to about half that level currently, while cotton production in the francophone region showed impressive gains, rising from 670,000 hectares in the 1980s to 2.4 million hectares currently. The devaluation of the CFA franc in 1994 gave a boost to cotton production in francophone Africa.

African cotton exports climbed from about 600,000 tons in the early 1980s to an estimated 1.6 million tons currently. Exports from North Africa are trending downward and are estimated at less than 200,000 tons this season. However, exports from Southern and East Africa are expanding to nearly 400,000 tons this season, and exports from the francophone region are forecast at a near record of 1million tons. Burkina Faso is the largest exporter from Africa, accounting for 300,000 tons in 2005/06; Mali is second, with 250,000 tons in 2005/06; and Benin, Cameroon and Egypt exported 100,000 tons or more. In total, 30 countries in Africa export some cotton while 37 countries are producing cotton. Egypt, Morocco and South Africa are the largest importers and combine to account for about 160,000 tons.

As was the case for exporters around the world, during 2005/06 the largest market for African cotton exports was China, which has taken between one-third and one-half of African exports in recent seasons. Indonesia, Bangladesh, Viet Nam, Chinese Taipei and Thailand are also significant markets for African exports. India has become an important importer in recent years, especially for fine cottons from Egypt.

TRENDS IN MILL USE OF FIBRE

World textile fibre consumption is driven by three major economic variables: income, population growth and fibre prices. World demand for textile fibres has increased at an impressive pace since the 1950s. From 7.6 million tons in 1950, textile fibre consumption increased to 56 million tons in 2004. While about 50% of the increase was the result of population growth, the remaining 50% was the result of higher income per capita, declines in real textile prices, and competition, which generated new uses for textile fibres. However, the rate of growth of fibre consumption has decelerated gradually. The average annual rate of growth of textile fibre consumption was 3.7% during the 1960s, 3.1% during the 1970s, 2.5% during the 1980s and 2.7% during the 1990s. Growth of the two major economic variables that determine textile consumption, income and population, decelerated during the 1990s compared with the 1960s.

An exogenous factor that has supported textile consumption in the last few years is the gradual integration of textile trade into World Trade Organization (WTO) rules. By December 2004, just over half of world textile trade had already been gradually integrated, and on 1 January 2005, all textile trade was integrated into WTO rules. Therefore, quotas agreed under the Multifibre Arrangement (MFA) no longer exist. Research by the International Cotton Advisory Committee (ICAC) Secretariat, using previous joint work with the Food and Agriculture Organization of the United Nations (FAO), suggests that because of textile quota elimination, the world was consuming half a million tons more cotton by the end of 2005. A large portion of the gains in cotton consumption due to quota elimination probably occurred between 1995 and 2004, particularly since January 1, 2002.

EXPANSION OF RETAIL COTTON USE

The rate of expansion of world cotton consumption accelerated for the fourth year in 2005, and did so while consumption of other fibres stagnated. World cotton consumption increased every year between 1998 and 2006.

Lower prices of cotton relative to other fibres, strong economic growth, and the popularity and greater availability of cotton products compared to products made of other fibres supported increases in cotton consumption. Contrary to the trend in the overall textile market, increases in world cotton consumption in 2005 were concentrated in developing countries. The world consumed an additional 1.8 million tons of cotton products in 2005, and of that additional consumption, developing countries accounted for 72%, industrialized countries for 26%, and Central and Eastern Europe and the former Soviet Union for 2%. As cotton consumption has increased in developing countries at higher rates in recent years, developing countries' share of the world end-use cotton market increased to more than 50% in 2005.

Relative cotton prices declined between 2003 and 2005, and in 2005 cotton prices were at their lowest level relative to polyester since 1992. Importantly, cotton prices have been less volatile in recent years than they were in the 1990s. Cotton prices, as measured by the Cotlook A Index, declined from an annual average of $0.63 per pound in 2003, to an average $0.62 per pound in 2004 and $0.55 per pound in 2005.

RETAIL COTTON CONSUMPTION

In 2003, developed countries as a group accounted for 44% of retail-level cotton consumption worldwide, and developing countries accounted for 52%. At the retail level, the United States is the largest consuming country, accounting for 21% of total cotton use in 2005. United States per capita cotton consumption was 17 kilograms in 2005, compared with a world average of only 3.8 kilograms. High consumer incomes, a history of cotton consumption, consumer preferences in favour of cotton bolstered by industry advertising, and fashion trends that favour cotton explain the high level of per capita cotton use in the United States.

Retail consumption of cotton in Latin America accounted for 9% of world cotton use in 2000; per capita consumption was 3.2 kilograms per year. Consumers in Brazil and Mexico account for two-thirds of Latin American retail-level cotton use.

Retail consumption in the EU-15 accounts for 16% of world cotton use, and per capita cotton consumption in Europe was about 7 kilograms in 2000. The lower level of per capita consumption of cotton in Europe compared with the United States reflects lower average income levels, fewer consumer-oriented retail structures, and differences in tastes and preferences between United States and European consumers.

Retail consumption in the Russian Federation and other countries of the former Soviet Union accounted for 2% of world cotton use in 2000; per capita cotton use was below the world average at just 2.7 kilograms.

Retail consumption in the Middle East, including Turkey, accounted for 6% of world use in 2000; per capita consumption was equal to the world average at 3.6 kilograms.

Africa, including South Africa and Egypt, accounts for only 2% of world cotton use at the retail level; per capita consumption of cotton in Africa is less than 1 kilogram per year.

Retail consumption in Japan equalled 6% of world cotton use in 2000. Per capita consumption in Japan was 9 kilograms, 2 kilograms higher than the EU average, but lower than in the United States. Consumption in the rest of East Asia (including China) and South Asia accounted for 31% of world cotton use at the retail level in 2000, but per capita consumption averaged just 1.8 kilograms because of low incomes and government policies that favour the use of polyester to conserve land devoted to cotton. One of the great challenges for the cotton industry is to raise per capita consumption in the countries with the largest populations, including China, where cotton use per capita was just 1.9 kilograms in 2000, India, with per capita cotton use of 1.7 kilograms, and Indonesia, with per capita use of 1.4 kilograms. It is hoped that rising incomes in India, Indonesia and China will lead to increases in per capita cotton consumption during the current decade.

MILL CONSUMPTION

World mill use of cotton rose from 6 million tons in the early 1950s to 26 million tons in the mid-2000s. The average rate of increase in world cotton mill use over the last six decades was about 290,000 tons per year. Mirroring end-use consumption, world mill consumption of cotton was stagnant during the first half of the 1990s, growing by only 0.6% between 1990 and 1997, but increased rapidly thereafter. In the early 1990s, mill consumption of cotton declined dramatically in Eastern Europe and the former Soviet Union, from 2.5 million tons in 1990/91 to 730,000 tons in 1998/99, offsetting gains elsewhere in the world. Mill consumption of cotton in the former COMECON group of countries (Council for Mutual Economic Cooperation), an economic grouping of the Soviet Union, Eastern Europe, Viet Nam and Cuba, recovered to 1 million tons in 2005/06. Mill consumption of cotton in industrialized countries remained at about 4 million tons during the early 1990s, but declined rapidly after 1998/99 to 1.8 million tons in 2005/06. High cost structures and increased import competition from developing countries caused the cotton textile industries in many industrialized countries to reduce production beginning in the 1990s.

Mill consumption of cotton in developing countries increased at an annual rate of 3.9%, from 8.5 million tons in 1980/81 to 12.3 million tons in 1990/91. Growth of mill consumption decelerated during the first seven years of the 1990s to an average annual rate of 2.7%, with annual consumption reaching 14.3 million tons in 1997/98, but regained strength since 1998/99, growing at an average annual rate of 6% to exceed 23 million tons in 2006/07. The bulk of the increase since 1998 occurred in China, but important expansions were also registered in India, Pakistan and Turkey. As a result, the processing of cotton continued to be concentrated in developing countries, and their share of world mill consumption rose from 67% in 1990/91 to 90% in 2006/07, compared to 46% in 1970/71 and 28% in 1950/51.

For the past eight years, China has been the driving force of the world textile industry. Between 1998/99 and 2006/07, the increase in mill consumption of cotton in China accounted for 84% of additional consumption worldwide. The Chinese industry processed 10.5 million tons of cotton in 2006/07, an increase of about 6 million tons since 1998/99. Chinese mill use of cotton in 2006/07 accounted for 40% of global mill use, up from 23% in 1998/99. The textile industry in China is highly dependent on the export market, and China has increased its share of world textile and apparel exports in the last eight years. During the 1990s, mill consumption of cotton became more concentrated in the largest processing countries. In 1980/81, the six countries that are the largest processors today (China, India, Pakistan, the United States, Turkey, and Brazil), accounted for 51% of world mill consumption. These countries accounted for 57% of world mill consumption in 1990/91, and 79% in 2006/07.

INTER-FIBRE COMPETITION

Cotton faces competition from chemical fibres. At the start of the twentieth century, cotton had a dominant share of the textile market. At the beginning of the twenty-first century, cotton is one of many fibres available and has been surpassed by polyester. Cotton consumption per capita has been almost constant since 1960, while total textile fibre consumption per capita has more than doubled.

World consumption of all textile fibres, including cotton, chemical fibres and wool, increased at an impressive pace, from 9.6 million tons in 1950 to 56 million tons in 2004. Fibres competing with cotton include natural fibres and chemical fibres, primarily polyester. Cotton's share of world textile fibre use fell from more than 70% in the 1950s to less than 50% by the end of the 1970s. Cotton did better in the 1980s, but its share of world textile fibre fell to about 40% by 2005.

Cotton's major advantages over its primary competitors in the chemical fibre complex include wearing comfort, natural appearance, moisture absorbency, its status as a renewable resource and the important economic role of cotton in many producing countries. However, cotton also suffers from several disadvantages in comparison with chemical fibres, including contamination introduced during harvest, ginning and handling, and annual fluctuations in the quantity and quality of production and consequent variability in prices. Cotton also has difficulty meeting the needs of modern spinning equipment for strength, uniformity and other quality parameters.

Cotton's share of world fibre use exceeded 60% in the 1960s, fell to 50% during the 1980s and dropped to less than 40% in the early 2000s. However, cotton experienced a revival in use during 2004 and 2005 linked to lower prices, and cotton's share of fibre use increased to above 40% in 2005. However, over the longer term, cotton is expected to continue to lose market share.

LONG-TERM PROJECTIONS FOR TEXTILE FIBRE CONSUMPTION

The set of assumptions used to project world textile fibre and cotton consumption in 2010 and 2020 includes:

- Annual average gross domestic product (GDP) growth since 1970 as a proxy for long-term GDP growth;
- Population projections from the United Nations;
- A long-term cotton price trend equal to average relative cotton prices between 2003 and 2005; and
- Increases in the ICAC Textile Fibre Price Index in tandem with inflation.

World textile fibre consumption is projected to expand at an annual average rate of 4% to reach 70 million tons in 2010 and by 2.8% per year to reach 87 million tons in 2020. Lower rates of growth in world textile fibre consumption aremainly associated with lower world GDP growth (down from 5.3% during the 1960s to 3.3% during the 1990s) and lower growth of the world population (from 2.1% during the 1960s to 1.7% during the 1990s). Between 2000 and 2005, the average rate of growth of textile fibre consumption was 3.8%. World cotton consumption is projected to expand at an annual average rate of 2% to reach 26.7 million tons in 2010 and 32 million tons in 2020. Cotton's share of the world textile fibre market is projected to decline to 37% in 2020.

TRADE

World trade in cotton rose from 2.6 million tons in 1950/51 to 4 million tons in the early 1970s and reached 5.8 million tons in 1986/87. Cotton exports averaged 5.9 million tons during the 1990s and climbed to a record of 9.7 million tons in 2005/06. Among the top seven cotton-producing countries, only Uzbekistan does not rank among the top seven consuming countries. Trade accounted for 40% of world cotton production in 2005/06, and the value of world exports was approximately $12 billion.

World trade in cotton is projected at 8 million tons in 2006/07. Production is falling behind mill use in China, Pakistan and Turkey. These three countries accounted for 15% of world imports in 2000/01 and for an estimated 40% in 2006/07, while imports by the rest of the world declined.

The largest and most significant impetus to the growth of world trade in cotton is provided by a sharp increase of cotton use in China. A record surge of cotton imports by China to 4.2 million tons, or 44% of world imports, in 2005/06, led world trade to a record. With the reduction of stocks in China to minimum levels, the Government began to provide full support to imports by issuing sufficient import quotas as a measure to balance supply and use, reduce domestic prices, and make the textile industry more competitive. For the fifth season in a row Turkey is the second-largest importer of cotton, accounting for 700,000 tons or 8% of world imports in 2006/07. Between 1998/99 and 2004/05, mill use in Turkey rose by 450,000 tons and reached 1.55 million tons. However, Turkey also faces competition in textile exports from Asian suppliers, and mill use in 2006/07 is estimated at the same level of 1.55 million tons. Because cotton production in Turkey remains behind increasing use, imports remain a significant source of supply.

India became one of the leading importers of cotton starting in 1999/2000 because of reduced production due to reduced planted area and drought. During the same period, Indian mill consumption was stable at around 2.9 million tons, supported by strong exports of cotton yarn and textile exports to Asian markets, the United States, Canada and Mexico. In 2001/02, India imported 520,000 tons of cotton, accounting for 8% of world imports. Indian area and yields rose during 2004/05, resulting in a crop of 3.9 million tons; as a result of increased domestic supply, imports by India declined to 150,000 tons. By 2005/06, production in India had begun to exceed mill use and instead of importing cotton, India exported 700,000 tons, the third-highest total in the world. Indian exports in 2006/07 are estimated at 960,000 tons.

Cotton consumption in Pakistan continues to expand rapidly in response to export-driven demand. Between 1998/99 and 2006/07, mill use in Pakistan rose by 6% a year to an estimated 2.6 million tons.

World cotton imports are projected to reach 9 million tons in 2007/08, the second highest level of imports since the record of 9.7 million tons was reached in 2005/06. Increased production in importing countries led to a decline in world imports to 8.2 million tons in 2006/07. China (Mainland) contributed the most to the increased supply and reduced world imports in 2006/07 as a result of a record domestic supply. China (Mainland), Turkey, Bangladesh, Indonesia, Pakistan and Thailand became the largest cotton importers during the past decade.

The largest share of increased world import demand is being met by exports from the United States. Large supplies of cotton in the United States and declining mill use led to record United States exports of 3.8 million tons in 2005/06. United States exports are estimated at 2.95 million tons in 2006/07 because of smaller production in the United States and reduced imports by China. The next largest exporters are India, Uzbekistan, Australia, West Africa and Brazil. Exports from Uzbekistan were mostly declining, from 1.3 million tons in 1992/93 to 660,000 tons in 2003/04. The reason for the steady decline in exports was a decline in production and increased mill use. In 2004/05 and 2005/06, production in Uzbekistan rebounded and exports increased to 1 million tons, accounting for 10% of world exports. Cotton area in Uzbekistan is projected to remain stable during the next several seasons. At the same time, Uzbekistan is expected to continue expansion of spinning capacity, increasing utilization of cotton domestically and reducing the availability of supplies for exports.

Between 1991/92 and 2002/03, Brazil was a net importer of cotton. During the past several seasons cotton production has begun to rise rapidly because of new, high-yielding commercial production in central Brazil, including the state of Mato Grosso. In 2003/04, cotton production in Brazil exceeded consumption by almost half a million tons, and exports by Brazil rose to 430,000 tons in 2005/06.

Exports from the CFA zone (francophone Africa) reached a record of more than 1 million tons in 2003/04 and exports were above 1 million tons again in 2005/06. The region is experiencing a reorganization of its cotton sector as several countries adopt policies encouraged by the World Bank and donor countries to encourage privatization. Production in francophone Africa may decline in coming years. Lower market prices for cotton, caused partly by subsidies paid in developing countries and partly by a weakening of the United States dollar against the CFA, are a particular source of dissatisfaction among African producers.

Australian production has suffered from severe drought during the past several seasons, leading to a sharp decline in exports. In 2004/05 Australian exports fell to 435,000 tons, compared with 850,000 tons in 2000/01. Exports from Australia are projected at 500,000 tons during 2006/07 as stocks are being drawn lower. Australia accounted for only 7% of world exports in 2004/05, compared with 14% during 2000/01.

THE IMPORTANCE OF COTTON IN WORLD TRADE

Cotton is essentially produced for its fibre, which is universally used as a textile raw material. Cotton is an important commodity in the world economy. Grown in more than 100 countries, cotton is a heavily traded agricultural commodity, with over 150 countries involved in exports or imports of cotton. The six largest consuming countries are also among the top seven producing countries. Cotton trade accounted for about 30% of world output from 1980/81 through 2004/05, but this share rose to almost 40% in 2005/06.

As world cotton production inevitably varies from year to year, variations in supply can cause wide fluctuations in price. The estimated nominal value of world exports dropped from a high of $13 billion in 1994/95 to $6 billion in 2001/02 and rebounded to approximately $12 billion in the 2005/06 marketing season (August-July). World cotton trade is not highly concentrated compared to other commodities. About 500 firms are involved in cotton exports worldwide.

World cotton exports reached 3.6 million tons in 1926/27, and that level was not surpassed until the early 1950s. Exports climbed to 9.8 million tons in 2005/06. The United States plays the leading role in cotton exports. Raw cotton was the United States' largest merchandise export from 1803 through 1937, and the United States has been the largest cotton exporter since 1834 (with only one exception in 1985/86).

Cotton is also a very political crop because of its importance in world trade and to the economies of many developing countries. In many countries, cotton exports not only are a vital contribution to foreign exchange earnings but also account for a significant proportion of GDP and tax income. Cotton is playing a major role in economic development in Africa: 37 of the 53 African countries produce cotton and 30 are exporters. Most Central Asian republics from the former Soviet Union are also very dependent on cotton exports.

Based on average export values in 2004/05, cotton was among the top three export products in 10 countries, and the average share of cotton exports in total product export earnings exceeded 10% in seven countries.

Nevertheless, cotton represents a very small share of world trade in terms of value. In UNCTAD export statistics by product, cotton ranked 170th on average 2004/05 values, accounting for 0.11% of world product exports in 2005 ($11.4 billion).

The cotton export market is relatively concentrated. With an index value of 0.386 in 2005, cotton is ranked twenty-first among all commodities according to the concentration index calculated by UNCTAD (an index value that is close to 1 indicates a very concentrated market; values closer to zero reflect a more equal distribution of market share among exporters or importers). The cotton import market is less concentrated. With a concentration index value of 0.294, cotton is ranked thirty-fourth among all imported commodities. However, the structural change index value of 0.378 in 2005 reveals a significant change in the composition of importers compared to the reference year (1995).

The ICAC Secretariat has studied the structure of world trade since 1994 and compiles a list of cotton-trading companies every year. The world cotton industry is not very concentrated by the standards of industrial markets and the international cotton shipping industry is highly competitive About 500 firms are engaged, at least in part, in international trade in cotton.

5

Prospects for Bt Cotton Technology

INTRODUCTION

Cotton provides a livelihood to more than 60 million people in India by way of support in agriculture, processing, and use of cotton in textiles. Cotton contributes 29.8% of the Indian agricultural gross domestic product, and nearly nine million hectares of land in India is used to produce 14.2 million bales of cotton lint.

Indian cotton production is third in the world in quantity, although the productivity is substantially low. The major reason for this low productivity is damage caused by insect pests—notably *Helicoverpa armigera*, commonly referred to as American Bollworm. Nearly Rs.12 billion worth of pesticides are used in India to control just the bollworm complex of cotton. Mahyco (Maharashtra Hybrid Seed Company), in collaboration with Monsanto, has introduced Bt cotton technology into India. Bt cotton carries the *Cry1Ac* gene derived from the common soil bacterium *Bacillus thuringiensis* var. *kurstaki*, which results in the expression of the *Cry1Ac* protein that confers resistance to the bollworm complex.

DESCRIPTION

Strains of the bacterium *Bacillus thuringiensis* produce over 200 different Bt toxins, each harmful to different insects. Most notably, Bt toxins are insecticidal to the larvae ofmoths and butterflies, beetles, cotton bollworms and ghtu flies but are harmless to other forms of life. The gene coding for Bt toxin has been inserted into cotton, causing it to produce this natural insecticide in its tissues. In many regions, the main pests in commercial cotton are lepidopteran larvae, which are killed by the Bt protein in the transgenic cotton they eat. This eliminates the need to use large amounts of broad-spectrum insecticides to kill lepidopteran pests (some of which have developed pyrethroid resistance). This spares natural insect predators in the farm ecology and further contributes to noninsecticide pest management. However, Bt cotton is ineffective against many cotton pests such as plant bugs, stink bugs, and aphids;

depending on circumstances it may be desirable to use insecticides in prevention. A 2006 study done by Cornell researchers, the Center for Chinese Agricultural Policy and the Chinese Academy of Science on Bt cotton farming in China found that after seven years these secondary pests that were normally controlled by pesticide had increased, necessitating the use of pesticides at similar levels to non-Bt cotton and causing less profit for farmers because of the extra expense of GM seeds.

MECHANISM

Bt cotton was created through the addition of genes encoding toxin crystals in the Cry group of endotoxin. When insects attack and eat the cotton plant the Cry toxins are dissolved due to the high pH level of the insects stomach. The dissolved and activated Cry molecules bond to cadherin-like proteins on cells comprising the brush border molecules. The epithelium of the brush border membranes separates the body cavity from the gut whilst allowing access for nutrients. The Cry toxin molecules attach themselves to specific locations on the cadherin-like proteins present on the epithelial cells of the midge and ion channels are formed which allow the flow of potassium.Regulation of potassium concentration is essential and, if left unchecked, causes death of cells. Due to the formation of Cry ion channels sufficient regulation of potassium ions is lost and results in the death of epithelial cells. The death of such cells creates gaps in the brush border membrane.

ADVANTAGES

Bt cotton has several advantages over non Bt cotton. The important advantages of Bt cotton are briefly pointed out as bellow:

- Increases yield of cotton due to effective control of three types of bollworms, viz. American, Spotted and Pink bollworms.
- Insects belonged to Lepidoptera (Bollworms) are sensitive to crystalline endotoxic protein produced by Bt gene which in turn protects cotton from bollworms.
- Reduction in pesticide use in the cultivation of Bt cotton in which bollworms are major pests.
- Reduction in the cost of cultivation and lower farming risks.
- Reduction in environmental pollution by the use of insecticides rarely.
- Bt cotton exhibit genetic resistance or inbuilt resistance which is a permanent type of resistance and not affected by environmental factors. Thus protects crop from of bollworms.
- Bt cotton is ecofriendly and does not have adverse effect on parasites, predators, beneficial insecticides and organisms present in soil.
- It promotes multiplication of parasites and predators which help in controlling the bollworms by feeding on larvae and eggs of bollworm.

- No health hazards due to rare use of insecticides (particularly who is engaged in spraying of insecticides).
- Bt cotton are early in maturing as compared to non Bt cotton.

DISADVANTAGES

Bt cotton has several advantages but it has some limitations also, which were given as below;

- High cost of Bt cotton seeds as compared to non Bt cotton seeds.
- Effectiveness up to 120 days, after that the toxin producing efficiency of the Bt gene drastically reduces.
- Adverse effect on insecticide manufacturing companies due to reduced use of pesticides.
- Adverse effect on the employment of those persons engaged in pesticide industries.
- Ineffective against sucking pests like jassids, aphids, whitefly etc.
- Potential adulterant malpractices such as mixing of lowcost non Bt cotton seeds with high cost Bt cotton seeds. The sale of this mixture could affect cotton production.

IN INDIA

Bt cotton is supplied in India's Maharashtra state by the agri-biotechnology company, Mahyco, as the distributor.

The use of Bt cotton in India has grown exponentially since its introduction. Recently India has become the number one global exporter of cotton and the second largest cotton producer in the world. India has bred Bt-cotton varieties such as *Bikaneri Nerma* and hybrids such as NHH-44, setting up India to benefit now and well into the future.

Socio-economic surveys confirm that Bt cotton continues to deliver significant and multiple agronomic, economic, environmental and welfare benefits to Indian farmers and society including halved insecticide requirements and a doubling of yields.

However, India's success has been subject to scrutiny. Monsanto's seeds are expensive and lose vigour after one generation, prompting the Indian Council of Agricultural Research to develop a cheaper Bt cotton variety with seeds that could be reused.

The cotton incorporated the cry1Ac gene from the soil bacterium Bacillus thuringiensis (Bt), making the cotton toxic to bollworms. This variety showed poor yield, was removed within a year, and contained a DNA sequence owned by Monsanto, prompting an investigation. In parts of India cases of acquired resistance against Bt cotton have occurred. Monsanto has admitted that the pink bollworm is resistant to first generation transgenic Bt cotton that expresses the single Bt gene (Cry1Ac). The state of Maharashtra has banned the sale and distribution of Bt cotton in 2012, to promote local Indian seeds, which demand less water, fertilizers and pesticide input.

ADVANTAGES

The main selling points of Bt cotton are the reductions in pesticides to be sprayed on a crop and the ecological benefits which stem from that. China first planted Bt cotton in 1997 specifically in response to an outbreak of cotton bollworm, *Helicoverpa armigera*, that farmers were struggling to control with conventional pesticides. Similarly in India and the US, Bt cotton initially alleviated the issues with pests whilst increasing yields and delivering higher profits for farmers.

Studies showed that the lower levels of pesticide being sprayed on the cotton crops promoted biodiversity by allowing non-target species like ladybirds, lacewings and spiders to become more abundant. Likewise it was found that integrated pest management strategies (IPM) were becoming more effective due to the lower levels of pesticide encouraging the growth of natural enemy populations.

CONTROVERSIES

Following the introduction of Bt cotton in northern China, non-target pests such as mirid bugs (*Heteroptera: Miridae*) have been becoming more abundant due to the reduction in conventional pesticides being sprayed. Following from this, a second issue being seen across the world, is the development of resistance genes in pest populations limiting the effect of Bt crops.

The main drivers for this widespread resistance in China and India included the high proportion of Bt cotton being planted, 95% and 90% respectively in 2011, and the low implementation of refuge areas.

Refuge areas act to limit the development of resistance genes in a targeted pest population. The Environmental Protection Agency (EPA) in the US requires farmers to have refuge areas of 20%-50% non-Bt crops within 0.8 km of their Bt fields. Such requirements were not seen in China, where instead farmers relied on natural refuge areas to decrease resistance.

A novel solution to the resistance problem was trialed in Arizona, with sterile male pink bollworms (*Pectinophora gossypiella*) being released into populations of their wild Bt-resistant counterparts. The hypothesis being tested was whether sterile males mating with the few surviving females, who had developed resistance, would lead to a reduced density of pests in the following generation. The result was a dramatic reduction in pink bollworms, with only two pink bollworm larvae being found by the third year of the study.

In India, Bt cotton has been enveloped in controversies due to its supposed links with seed monopolies and farmer suicides. However, the link between the introduction of Bt cotton to India and a surge in farmer suicides has been refuted by other studies, with farmer suicides actually having fallen since the introduction of Bt cotton according to some studies. Bt cotton accounts for 93% of cotton grown in India.

THE NEED FOR BT COTTON

The genetic resistance, one of the important pest management strategy, is available in cotton gene pool against the sap sucking pests such as jassids, whitefly etc and using this several resistant / tolerant varieties and hybrids have been developed and released in India.

However, such kind of known resistance is not available against the bollworms. Hence, an alternate strategy is explored to circumvent this problem by cloning and transferring the genes encoding the toxic crystal ä - endo toxin protein from the soil bacterium *Bacillus thuringiensis*. The Bt transgenic cotton (Bollgard of Monsanto) has thus been developed successfully in USA, which has the ability to control the bollworms at the early stages of crop growth (upto 90 days) effectively.

The first commercial Bt cotton variety was released in USA by M/S. Monsanto (Bollgard), which contains Cry 1Ac gene of *Bacillus thuringiensis*. Bt cotton is commercially grown in several countries like China, Australia, Mexico, South Africa, Argentina, India, Indonesia etc.

World wide the area under Bt cotton keep increasing year by year. Overall, about 12% of the world cotton is now planted with Genetically Modified varieties / hybrids (GMO) and ICAC has estimated that his may rise to 50 % in 5-7 years.

CHRONOLOGY OF BT COTTON IN INDIA

March 10, 1995: Department of Biotechnology (DBT) of the Government of India permits import of 100 gm of transgenic Cocker-312 variety of cottonseed cultivated in the United States by Mahyco. This variety contained the Cry 1 Ac gene from the bacterium Bacillus thuringiensis.

April 1998: Monsanto-Mahyco tie up. Monsanto given permission for small trials of Bt cotton 100 g per trial by Department of Biotechnology (DBT).

January 8, 1999: RCGM expresses satisfaction over the trial results at 40 locations and on April 12th directs MAHYCO to submit applications for trials at 10 locations before MEC.

2000-2002: ICAR trials were conducted at different AICCIP centres of Central and South Zone locations.

February 20, 2002: The Indian Council of Agricultural Research (ICAR) submits a positive report to the Ministry of Environment on the field trials of Bt cotton. It is now expected that the Genetic Engineering and Approval Committee (GEAC) of the environment ministry will approve commerical use of Bt cotton within a month.

March 25, 2002: Approval given for commercial cultivation to three Bt Cotton hybrids of M/s. MAHYCO by GEAC

In India, the area under Bt cotton hybrid has increased from a mere 29,307 ha during 2002 to 12,50,833 ha in 2005. During 2006, the area has increased upto 30,00,000 ha as projected by seed industry. The state wise break up of area covered under Bt hybrids. It is estimated that roughly 40% of total area under cotton hybrids (approximately 100 lakh acres) can potentially be sown under Bt version, which translates into 40 lakh packets every year worth Rs. 640 crore. So far, 52 different Bt cotton hybrids were released for cultivation in India .

PERFORMANCE OF RELEASED BT COTTON HYBRIDS IN ICAR TRIALS

Comparative performance of Bt cotton hybrids developed by different private companies over the year (2001-2005) over different locations of three cotton growing zones of the country indicated superiority of Bt hybrids over their non-Bt counterparts in terms of both yield and contributing characters. Apart from seed cotton yield, they were also superior in number of bolls/plant, boll weight, seed index, lint index and ginning outturn.

The data on fibre quality of these hybrids indicated that there is no significant difference between Bt and non-Bt hybrids for fibre quality parameters *viz.*, 2.5 % span length [mm], bundle strength [g/tex] at 3.2 mm gauge, and Micronaire. When we compare the fibre quality of these hybrids as against the requirements of the mills, as prescribed by South India Mills Association, most of the hybrids are spinnable between 30s and 40s counts.

EVALUATION IN MULTILOCATION AND LARGE SCALE FIELD TRIALS

Large scale field trial results of four years from 1998 to 2002 indicated that Bt cotton was able to resist bollworm infestations thereby resulting in good boll retention and higher yields. Apart from the increase in yields there was a concomitant reduction in the use of insecticides due to Bt-cotton. Thus it was concluded that Bt-cotton has potential to improve the lives of cotton farmers through the provision of favourable environmental and economic consequences.

BT COTTON VIS-À-VIS ENVIRONMENTAL PROTECTION

Bt cotton cultivars exhibited excellent control of *Helicoverpa zea* and *Heliothis virescens* and reduced impact of insecticides to create eco-friendly environment without compromising yield. Since the Bt gene is effective during the early phase of crop growth, the bolls produced in the bottom most branches of the plant are retained fully. The lint obtained from the bottom 1/3rd part of the plant is reported to be of highest quality leading to the production of more quality fibres. Because of the retention of early formed bolls in the plant, the crop enters into senescence early and matures early compared to non-Bt counterparts. This ultimately helps in harvesting of seed cotton in two pickings.

The major emphasis was given to the control of boll worms in Bt hybrids as against their non-Bt counterparts by considering the number of times the Economic Threshold Level (ETL) crossed, total number of sprays given for the control of various insect pests under protected and unprotected conditions etc. In all these aspect, the released Bt cotton hybrids were found to be more efficient as compared to their non-Bt counterparts. As compared to insecticide control of bollworms, Bt cotton technology will not harm non-target beneficial insects, reduction in production cost, increased profit, reduced farming risk and improved economic outlook for cotton. Use of this technology is also helpful in improving wild life population, reduced run off insecticides, reduced air pollution and improved safety to farm workers and neighbourhood.

REDUCED BOLLWORM DAMAGE AND REDUCTION IN INSECTICIDE SPRAYS

Almost all the data available in India show that the Bt cotton hybrids crossed ETL for bollworm population in some locations only once after 90 days. Whereas the non-Bt and check hybrids crossed ETL more than three times at different locations from 60 DAS. The population of bollworms especially *Helicoverpa armigera* was significantly lesser in number in Bt cotton hybrids as compared to non-Bt and check hybrids. Amongst Bt hybrids, MECH-184 and MECH-12 recorded significantly lesser boll and locule damage compared to non-Bt and check hybrids, indicating that the two Bt hybrids exhibited higher tolerance to bollworm damage. There is a 50% overall reduction in the *H. armigera* larval population in Bollgard-MECH-162 compared to the non-Bt MECH-162. Bollgard-MECH-12, Bollgard-MECH-162 and Bollgard-MECH-184 were able to reduce larval populations of spotted bollworm (*Earias vittella*) up to 30-40% and pink bollworm (*Pectinophora gossypiella*) up to 60-80% in south India.

Large scale cultivation of Bt cotton has resulted in the significant reduction of insecticide use to the tune of 40 to 60% less than the intensity on the corresponding non-transgenic varieties. Several studies have evaluated the economic benefits accrued due to the cultivation of Bt transgenic cotton versus the corresponding non-transgenic cultivar. Apart from causing a reduction in the usage of insecticides all over the world Bt-cotton has significantly contributed to enhanced yields. Hence it has become very popular in all cotton growing countries of the world. One important advantage of Bt-cotton is that farmers rarely resort to prophylactic spray applications, which they do otherwise, in the absence of Bt-technology. In some regions of the country 7-

10 prophylactic sprays per season are common on cotton. The total number of sprays averaged at 16-20 in some districts of Andhra Pradesh and Punjab during 1986-2001. The problem of pest management had become more complex due to bollworm resistance to insecticides, thereby causing enormous wastage of insecticide and subsequent environment pollution.

Yield increased substantially by adopting Bt-cotton and farmers in India were able to reduce at least 2-3 insecticide applications. Over the past four years bollworm infestation in India was low, thus reducing the need for insecticide applications. However, the benefits of Bt-cotton were more in other countries where bollworm infestation was high. Insecticide applications on Bt-cotton varieties were reduced up to 14 applications in China, 7 in South Africa and 5-6 in Indonesia and Australia.

BIOSAFETY TESTS AND ASSESSMENT OF TOXICITY TO NON-TARGET ORGANISMS

Biosafety tests indicated absolute safety to goats, cows, buffaloes, fish and poultry. Feed-safety studies with Bt cottonseed meal were carried out with goats, buffalos, cows, rabbits, birds and fish. The results revealed that the animals fed with Bt-cotton seed meal were comparable to the control animals in various tests and showed no ill-effects. These studies were carried out by the National Dairy Research Institute, Karnal; Central Avian Research Institute, Bareily; Industrial Toxicological Research Centre, Lucknow; National Institute of Nutrition, Hyderabad; Central Institute of Fisheries Education, Mumbai and GB Pant University for Agriculture and Technology, Pantnagar.

The Cry1Ac is mainly toxic to the bollworms (cotton bollworm, pink bollworm and spotted bollworm), semiloopers and hairy caterpillars. Bt-cotton expressing Cry1Ac is absolutely non-toxic to all other non-target organisms such as beneficial insects, birds, fish, animals and human beings. Laboratory and field studies carried out in India showed that the Cry1Ac protein deployed in Bt-cotton did not have any direct effect on any of the non-target beneficial insects. Work carried out elsewhere in the world also showed similar results. There was some evidence of a reduction in numbers of predators and parasitoids which specialise on the Bt controlled bollworms, but also of increases in numbers and diversity of generalist predators such as spiders. Generally the decrease in the parasitoid and predator populations were associated with decrease in the densities of the pest populations on account of Bt-cotton. Due to these changes in pest complex, farmers had to spray 3-5 times on bollgard as compared to 6-8 times on non-Bt cottons. Any effects could be assigned to the decrease in prey quality – for example with stunted *Spodoptera litura* caterpillars which had fed on Bt cotton. In the field situation, partial life studies broadly confirmed this finding. There was no increase in green vegetable bug numbers, aphid or whitefly numbers on Bt cotton. In general, such adverse effects as have been measured are very small when compared with the side effects of the spraying of conventional insecticides.

POST RELEASE FIELD PERFORMANCE AND ENHANCED ECONOMIC BENEFITS

Results from extensive Bt cotton trials under farmer field conditions, conducted from 1998 to 2001 confirmed that Bt cotton with the Cry1 Ac gene provides effective and safe control of bollworm and related pests. Field trials have confirmed that, compared to conventional hybrids, Bt cotton can increase yields by up to at least 40%, reduce insecticide sprays by at least 50 % or more (decrease from 7 to 2 or 3 sprays on average) equivalent to savings of Rs 2500/ hectare, and increase overall farmer income from Bt cotton from Rs 3500 to Rs 10,000 or more per hectare. Mahyco commissioned a nationwide survey by ACNeilsen-ORG MARG in 2003. The survey covered 3,063 from Maharashtra, Madhya Pradesh, Andhra Pradesh, Karnataka and Gujarat.

The data showed that a yield increase by about 29% (range 18 to 40%) due to effective control of bollworms, a reduction in chemical sprays by 60% (range 51 to 71%) and an increase in net profit by 78% (range 66 to 164%) as compared to non-Bt cotton. The net profit was estimated to an average of Rs.7,724 (range Rs. 5,900 to 12,696) per hectare. Mahyco conducted an independent survey during 2003 to assess the performance of Bt-cotton in fields of 3000 farmers. Results showed an average net profit of Rs 18,325 (range 15,854 to 20,196) per hectare. Field trials conducted by Mahyco during 2001, in 157 farms in 25 districts of Madhya Pradesh, Maharashtra and Tamilnadu showed that there were no changes in the insecticide use for sucking pest control, but at least three sprayings meant for bollworm control were saved due to the Bt-technology. Thus, insecticide use of cotton bollworm was reported to have been reduced by 83% and yield increase by a staggering 80%. Global estimates show that, Bt cotton caused an average net income increase of $ US 50/hectare in the USA, $357/hectare to $549/hectare in China and $25-51/hectare in South Africa.

OPPORTUNITIES FOR THE FUTURE

Table. Area, Production and Productivity of cotton in India

Year	Area (lakh ha)	Production (lakh bales)	Productivity (kg lint/ha)
2000-01	81.5	167	319
2001-02	85.9	153	309
2002-03	73.9	158	322
2003-04	76.3	179	399
2004-05	89.2	243	463
2005-06	88.2	243	465

Table. Bt-cotton area (ha) in India, based on the number of packets (450 g) sold

S. No.	State	2002	2003	2004	2005
1	Andhra Pradesh	3,792	5,199	73,890	226,684
2	Madhya Pradesh	1,482	12,968	87,894	142,062
3	Gujarat	9,104	42,097	134,034	147,335
4	Maharashtra	12,379	18,711	208,715	621,111
5	Karnataka	2,178	3,547	20,443	28,888
6	Tamilnadu	373	3,404	9,756	18,409
7	Haryana	0	0	0	13,309
8	Punjab	0	0	0	51,425
9	Rajasthan	0	0	0	1,610
Total		**29,307**	**85,927**	**53,4731**	**1250,833**

Table. Area under Bt cotton hybrids in 2006-07 (Lakh hectares)

State	Total State Area	Bt cotton hybrid Area	% of Bt cotton Area
Punjab	6.18	2.81	45.5
Haryana	5.33	0.42	7.9
Rajasthan	3.08	0.05	1.6
Gujarat	23.90	4.07	17.0
Maharashtra	31.24	16.55	53.0
Madhya Pradesh	6.66	3.02	45.3
Andhra Pradesh	9.48	6.57	69.3
Karnataka	3.56	0.80	22.5
Tamil Nadu	0.94	0.32	34.0
Total	91.37	34.61	37.9

Keeping in mind the development of resistance to Cry 1Ac protein in insects, notable progress has already been made to diversify the transgene and to pyramid genes which are having different mode of action so that development of resistance is delayed.

The genes available for exploitation include Bollgard II of Monsanto (Cry 1Ac + Cry 2Ab), VIP COT of Syngenta (VIP 3A) and Wide Strike of Dow Agro Sciences (Cry 1Ac + Cry 1F). The problem with regard to fibre quality of Bt hybrid as noted in MECH. 12 and MECH. 184 may be circumvented by involving a proper combination of parents to produce superior Bt cotton hybrid for both yield as well as fibre quality.

Table. List of Bt cotton hybrids approved for commercial cultivation in India.

Name of the company	North Zone	Central Zone	South Zone
M/s. MAHYCO Seeds	MRC 6301 [2005] MRC 6304 [2005] MRC 6025 (2006) MRC 6029 (2006)	MECH. 12 [(2002], MECH. 162 [2002], MECH. 184 [2002], MRC 6301 [2005] MRC 7301 BG II (2006) MRC 7326 BG II (2006) MRC 7347 BG II (2006)	MECH. 12 (2002) MECH. 162 (2002) MECH. 184 (2002) MRC 6918 (HxB) (2005) MRC 6322 (2005) MRC 7351 BG II(2006) MRC 7201 BG II (2006)
M/s. Rasi Seeds	RCH 134 [2005] RCH 317 [2005] RCH 308 (2006) RCH 314 (2006)	RCH 2 [2004], RCH 118 [2005] RCH 138 [2005], RCH 144 [2005] RCH 377 (2006)	RCH 2 (2004) RCH 20 (2005) RCH 368 (2005) RCH 111 (2006) RCH 371(2006) RCHB 708 (HxB) (2006)
M/s. Ankur Seeds	Ankur 651 [2005] Ankur 2534 [2005]	Ankur 651 [2005] Ankur 09 [2005]	--------
M/s. Nuziveedu Seeds	NCS 913 (2006) NCS 138 (2006)	Bunny [2005] Mallika [2005] NCS 913 (2006)	Bunny (2005) Mallika (2005) NCS 913 (2006)
M/S Ganga Kaveri Seeds	--------	GK 204 (2006) GK 205 (2006)	GK 209 (2006) GK 207 (2006)
M/S Emergent genetics	--------	Brahma (2006)	Brahma (2006)
M/s Nath Seeds	NCEH 6 (2006)	NCEH 2 (2006)	NCEH 3 (2006)
JK Seeds	JKCH 1947 (2006)	JK Varun (2006)	JK Durga (2006) JKCH 99 (2006)
Ajeet Seeds	--------	ACH 33-1 (2006) ACH 155-1(2006) ACH 11-2 BG II (2006)	ACH 33-1(2006) ACH 155-1 (2006)
Prabhat Seeds Ltd	--------	NPH 2171 (2006)	PCH 2270 (2006) NPH 2171 (2006)
Krishidhan Seeds	--------	KDCHH 441 BG II (2006) KDCHH 9810 (2006) KDCHH 9632 (2006) KDCHH 9821 (2006)	KDCHH 9632 (2006) KDCHH 9810 (2006)
Vikram Seeds	--------	VICH 5 (2006) VICH 9 (2006)	VICH 5 (2006) VICH 9 (2006)
Tulasi Seeds	--------	Tulasi 4 (2006) Tulasi 117 (2006)	Tulasi 4 (2006) Tulasi 117 (2006)
Vikki's Agro Tech	--------	VCH 111 (2006)	--------
Pravardhan Seeds	--------	PRCH 102 (2006)	--------

Table. Economics of Bt-cotton cultivation in ICAR trials 2001.

Hybird	Yield Q/ha	Gross income Rs./ha	Insecticide cost Rs./ha	Net income Rs./ha
MECH-12 Bt	11.67	21,006	1,727	16,854
MECH-12 Bt	13.67	24,606	1,413	20,768
MECH-12 Bt	14.00	25,200	1,413	21,362
Local check	8.37	15,066	2,845	12,221
National check	7.31	13,158	2,001	11,157

Table. Spray application reduction on Bt cotton during 2002-2003

Country	Year of introduction	Total cotton area (lakh ha)	Area under Bt-cotton		No. of insecticide sprays	
			Lakh ha	%	Non-Bt	Bt-cotton
USA	1996	62.0	20.0	33	5	2
Mexico	1996	0.8	0.3	35	4	2
China	1997	48.0	15.0	31	20	7
Australia	1997	4.0	1.5	36	11	6
Argentine	1998	1.7	0.1	5	5	2
S. Africa	1998	0.4	0.2	45	11	4
Indonesia	2001	0.2	0.1	18	9	3
Colombia	2002	0.4	0.1	10	6	2
India	2002	85.0	2.8	3	5	2

Table. Results of the ACNeilsen-ORG MARG, 2003

State	Pesticide reduction		Yield increase		Net profit	
	Rs/ha	%	Q/ha	%	Rs./ha	%
Andhra Pradesh	4,594	58	4.9	24	12,717	92
Karnataka	2,930	51	3.3	31	6,222	120
Maharashtra	2,591	71	3.6	26	5,910	66
Gujarat	3,445	70	2.9	18	8,564	164
Madhya Pradesh	2,200	52	5.4	40	9,594	68
Average	3,202	60	4.2	29	7,737	78

BT COTTON IN INDIA -PROSPECTS AND PROBLEMS

Cotton the "White Gold" is an important cash crop of India which plays a vital role in the Indian economy. As an industrial crop, it supports millions of people through cultivation, processing and trade and contributes Rs.360 billion to the export income.

The area occupied by cotton in recent years fluctuated between 8 and 9 million hectares in India. For over three thousand years India was recognized as the cradle of cotton industry. India thus enjoys the distinction of being the earliest country in the world to domesticate cotton and utilize its fibre to manufacture fabric. The cultivation, processing and trade industries support millions of people and contribute 40 to 45% of the export income.

The cotton production reached an all time high of 270 lakh bales and surpassed the production of US to rank second in the world. One of the reasons attributed for this growth is introduction of Bt cotton in India. Currently Bt cotton occupies an area of 37-40% of total cotton area in India. Globally nine countries are growing Bt cotton. Two developed countries (USA and Australia) and seven developing countries including three Asian countries (China, India and Indonesia) three from Latin America (Mexico, Argentina and Columbia) and South Africa. Bt cotton was commercialized in India in the year 2002.

Table: Area under Bt cotton hybrids in 2006 – 07 (Lakh hectares)

State	Total State Area	Bt cotton hybrid area	% of Bt cotton area
Punjab	6.18	2.81	45.5
Haryana	5.33	0.42	7.9
Rajasthan	3.08	0.05	1.6
Gujarat	23.90	4.07	17.0
Maharashtra	31.24	16.55	53.0
Madhya Pradesh	6.66	3.02	45.3
Andhra Pradesh	9.48	6.57	69.3
Karnataka	3.56	0.80	22.5
Tamil Nadu	0.94	0.32	34.0
Total	**91.37**	**34.61**	**37.9**

COTTON BOLLWORMS

The cotton bollworm complex in India includes the false American bollworm (*Helicoverpa armigira*), Pink bollworm (*Pectinophora gossypiella*), spotted bollworm (*Earias vittella*) and spiny bollworm ((*Earias insulana*). Among the above *H.armigera* is the most dominant and difficult to control chiefly due to its wide spread insecticide resistance, multivoltine and prolific pattern of feeding and high polyphagy. It is highly destructive and wasteful feeder in the sense that a single larva can damage squares and bolls and had a wide distribution.

Chemical insecticides are used extensively on cotton crop for control of insect pests, especially bollworms. The number of sprays per crop season may vary from 5 to 20 or even more.

It is estimated that insecticides worth about Rs.30 billion are used annually in Indian Agriculture, of which, about Rs.16 billion are spent for the control of cotton pests and of this Rs.12 billion against boll worms alone. In terms of volume, about 54% of the total insecticides used in Indian Agriculture are sprayed on cotton crop. This indicates the economic importance of boll worms in general and *H.armigera* in particular.

Despite such huge efforts, bollworm control has not been generally satisfactory mainly because a pest like *H.armigera* has developed resistance to most of the currently recommended insecticides. Nevertheless farmers continue to use insecticides repeatedly as they have no option expect to 'spray' or 'pray'. This had frustrated the farmers, scientists and policy makers alike. Bt. Cotton came at a time when they were desperately looking an alternative and dependable control measure.

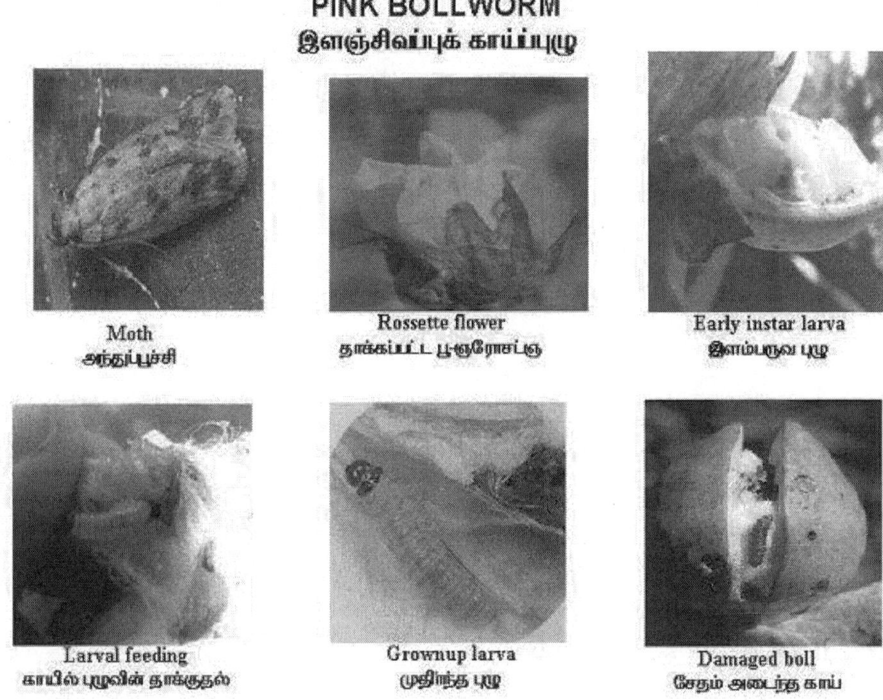

DEVELOPMENT OF BT – COTTON

The organism *Bacillus thuringiensis* was discovered in 1911 as a pathogen in flour moth, Thuringia, Germany. But later it was commercially utilized as biopesticides in France in 1938 and then in USA in 1950 for the toxin produced by this bacterium. From 1950 onwards the biopesticides containing this soil bacterium were popular. In 1992 the gene which is responsible for the toxin production was introduced to the cotton crop was grown in six locations in USA.

1996	-	The crop was economically grown in USA in 73000 ha
1997	-	In China 1 million ha
1998	-	1.5 million ha (USA, Mexico, Australia, Argentina, China and South Africa)
2001	-	5 million farmers grown Bt-cotton in which 99% were in developed countries
2002	-	Commercialization of Monsanto's Bt-cotton was approved
2003	-	Area increased approximately to 100,000 hectares in India

Presently 9 countries grow Bt-cotton. Two developed countries (USA and Australia) and seven developing countries including three Asian countries (China, India and Indonesia) three from Latin America (Mexico, Argentina, Colombia) and South Africa (Barwale *et al.*, 2004).

GENETICS OF BT-COTTON

The bacteria *Bacillus thuringiensis* produce two types of toxins namely 1. Cry (Crystal) toxin encoded by different cry genes and 2. Kytolytic toxin. Over 50 genes have been noted to encode for cry toxin and they are sequenced for various studies. The various genes and their properties are tabulated here under.

Gene	Crystal shape	Protein size (KD)	Insect actively
Cry I A(a), A(b), A(c), B, C, D, E, F, G	Bipyramidal	130 – 138	Lepidopteran larva
Cry II (Sub group) A, B, C	Cuboidal	69 - 71	Lepidoptera, diptera
Cry III (Sub group) A, B, C	Flat irregular	73 – 74	Coleoptara
Cry IV (Sub group) A, B, C, D	Biopyrimidal	73 – 134	Diptera
Cry V – IX	Various	35 – 129	Various

The commercial Bt–cotton available today contain genes from the isolate *B-thuringiensis,* ssp. *Kurstaki*that produces Cry I A(a), Cry I A(b), Cry I A(c), Cry IIA.

MODE OF ACTION OF THE TOXIN

In plants the toxin is present in non-toxic (protoxin) form, which is insoluble at low pH. So the protein is not toxic to human beings and higher organisms and present in the protein size of 130 – 138 KD.

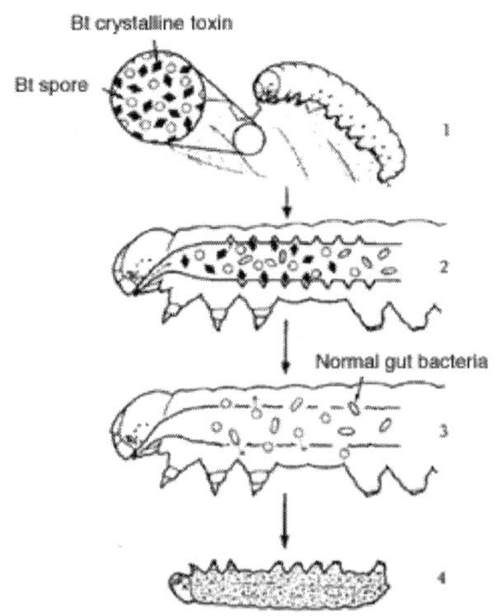

Whereas the insect mid gut is having pH of >9 which is suitable for the protoxin to dissolve and change to toxic form of size 60 KD.When cry proteins are ingested by insects, they are dissolved in the alkaline juices present in the mid gut lumen. The gut proteases process them hydrolytically to release the core toxic fragments.

The toxic fragments are believed to bind to a specific high – affinity receptors present in the brush – border of mid gut epithelial cells. As a result, the brush border membranes develop pores, most likely non-specific in nature which causes their swelling of mid gut and larva stops eating. Due to which pH is lowered and bacterial spore starts germinating which causes the death of larva called *septocaemia*.

USE OF BT-GENES IN COTTON

The Bt genes that are currently deployed are from two sources. Monsanto developed and deployed the Cry IA(c) gene in its Bollgard varieties, which are the most widely used in all nine countries that grow cotton. The second source is the Bt-fused gene that was developed by the public sector, Chinese Academy of Agricultural Sciences (CAAS) in Beijing, China (Jayaraman *et al.*, 2005)

Table. Genes utilized for the development of transgenic cotton hybrids in India

Company's Name	Gene utilised
Mahyco	cry1Ac
Monsanto	cry1Ac+2Ab
Nath seeds	cry1Ab+Ac fusion (China)
JK seeds	cry1Ac modified (IIT Khargpur)
Syngenta	vip3A +cry1 Ab
Dow Agri. Science	cry1 Ac + cry 1F,
Metahelix	cry1 C
ICAR	cry1 Aa3
	cry1F
	cry1 Ia5
	cry1 Ab
	cry1 Ac
NBRI	cry1 Ec

FIRST GENERATION BT-COTTON

The most prevalent Bt-gene on a global basis Cry I A(c) was incorporated into Coker 312 cotton designated MON 531 by Monsanto and later named Bollgard cotton (first generation Bt-cotton). The high transformation efficiency was achieved in Coker 312 with *Agrobacteium tumefaciens.* The transformed Coker was than back crossed with lines from Delta and Pine land and other companies that had necessary agronomic qualities for commercial acceptance.

The advantages of the Cry IA (c) in bollgard over the Bt-cotton spray are as follows:

1. Active protein expressed in all plant parts.
2. Active protein expressed throughout the season, hence timing of insecticide applications in relation to an infestation is not an issue.
3. Less farmer exposure to insecticide.
4. Labour saving technology due to elimination or reduction of insecticide sprays.
5. Contribution to and provides the foundation for an IPM strategy.

SECOND GENERATION BT-COTTON (BOLLGARD II COTTON)

The Insect Resistance Management (IRM) strategy for Bt-cotton that Monsanto in conjunction with USDA developed second generation of improved Bt-cotton with two Bt-genes, now designated Bollgard II. The new product bollgard II, Event 15985 was developed using particle acceleration plant transformation procedures to add the Cry II A(b) gene to the cotton line DP 50B that already had the cry 1 A(c) gene. Hence the bollgard II cotton contains Cry 1 A(c) and Cry II A(b). The dual gene cultivars are expected to provide growers with a broader control over a wide variety of insects.

In 2002, Dow Agrosciences announced the development of new Bt-cotton with traits that confer broad spectrum resistance to lepidopteran pest of cotton. The new Bt cotton product contains the dual genes Cry IA(c) and Cry IF, transformed with *Agrobacterium tumefaciens* and incorporated through back crossing into several high quality commercial varieties of cotton.

BT COTTON (BOLLGARD) IN INDIA

Realizing the importance of Bt cotton MAHYCO took the initiate introducing this technology into India in collaboration with Monsanto company.

As per regulatory procedure, Mahyco sent its application to Dept. of Biotechnology (DBT), Govt. of India, in March 1995 seeking permission for introducing this technology. On obtaining approval, Mahyco received 100 gms of Bt-cotton seeds (variety Cocker 312) containing the Bollgard Bt gene Cry I A(c) form Monsanto, USA, in March 1996. After testing the efficacy of this gene, these were used in breeding programmes and 40 elite Indian parental lines were introgressed with Cry I A(c) gene by crossing with Bt gene donar parent obtained from Monsanto. Some of the ruling conventional hybrids were converted into Bollgard using the converted parental lines and tested for their performance and three hybrids were released for commercial cultivation in March 2002 (MECH – 12, MECH – 162 and MECH – 184). This approval was proceeded by a large number of laboratory studies and about 500 field trials carried out during 1996 – 2001 to demonstrate the safety an benefit of Bt cotton as per regulatory requirements (Mohan and Manjunath, 2002).

Table. Development of Bt. Cotton in India

1995	Mahyco sent its application to Dept. of Biotechnology (DBT) seeking permission for introducing this technology.
1996	Mahyco received 100 gms of Bt-cotton seeds (variety Cocker 312) containing the Bollgard Bt gene Cry I A(c) form Monsanto, USA
1996 - 2002	After testing the efficacy of this gene, these were used in breeding programmes and 40 elite Indian parental lines were introgressed with Cry I A(c) gene by crossing with Bt gene donar parent Demonstration the safety and benefit of Bt cotton as per regulatory requirements.
2002 - 2003	Three Bt. Varieties namely, MECH 12, MECH 162 and MECH 184 were released commercially by Mahyco seed company. 3% cotton area under Bt. Cotton
2003- 2004	Bt. Varieties failure report from Andhra Pradesh
2004 - 2005	20 Bt. Strains released for commercial cultivation by various seed companies

EVALUATION OF BT COTTON

There is a 3 Tier Mechanism for Evaluation of Transgenics in India

RDAC: Review development in biotech at national and international levels, recommend safety

regulation in India

1. IBSC: Institute Biosafety Committee
2. RCGM: Review Committee on Genetic Manipulation
3. GEAC: Genetic Engineering Approval committee

SBCC: State Biotech Co-ordinate committee

DLC: District Level committee

Bt-cotton being a transgenic crop requires environmental clearance under rule 7-10 of the 1989"Rules for manufacture, Use, Import, Export and storage of hazardous microorganisms / Genetically Engineered Organisms or Cells" notified under the Environment (Protection) Act, 1986.

In India, two union ministries are involved in the regulation of GMO's - Ministry of Science and Technology (MoST) and Ministry of Environment and Forest. The Department of Biotechnology (DBT) functions under MoST. Two important committees namely Institutional Bio-safety Committee (IBSC) and Review Committee on Genetic modification (RCGM) work under the guidance of DBT. Another Major committee, namely Genetic Engineering Approval Committee (GEAC) is constituted under MoEF. These committees are represented by experts drawn from various fields and organization across the country and are responsible to ensure that proactive safety studies are carried out on GM products before they are approved for commercialization.

As per the direction and guidelines of the regulatory authorities, Mahyco carried out number of studies which includes molecular characterization of induced gene, biochemical characterization of expressed protein and Bt-protein expressed in Bt-cotton plants with regard to its potential for allergenicity, toxicity to non target organisms, gene flow and cross pollination, impact on soil microorganisms and food and feed safety evaluation of Bt cotton compared to non Bt-cotton seed.

INDIRECT CONSEQUENCES OF BT-COTTON

1. Insecticide use decline;
2. Reduced pyrethroids: Pink bollworm, mirid bugs & *Spodoptera* are re-emerging*H. armigera* & whitefly are taking a backseat
3. Parawilt: Some genotypes struggle with high boll retention, water imbalance, punctatum blood of hirsutum, undesirable heterosis (Mayee *et al.*, 2004)
4. High expectations, less results from rainfed and less fertile soils
5. Susceptibility to sucking pests –Linkage drag!!

DIFFERENCE BETWEEN BT. AND NON BT COTTON

The results demonstrated the following benefits of Bt cotton.

1. Good control of bollworm species
2. Significantly higher boll retention and more yield than the control
3. Reduction in chemical sprays for bollworm control (50% less than that required for conventional commercial hybrids.
4. Induces earliness about 20 – 30 days than the non Bt.
5. Substantial increase in net income to farmers.

6. No adverse impact on non-target organisms and the adjacent non Bt cotton or other crops.

According to survey carried out by ACNEILSEN - ORG MARG in 2003, over 90% of bollgard users and over 40% of non-users expressed the intent to purchase bollgard seeds in the next season (Manjunath, 2004).

Table. Hybrids released for South zone

Year	2006	2007
BG I intra hirsutem hybrids	22	27
BGII intra hirsutum hybrids	2	8
BG I inter specific hybrids	2	5

It is believed that perhaps the greatest challenge to the acceptance of Bt. cotton by the society would not be with the plants themselves but by the high expectations from the farmers if it is promoted unduly by making exaggerated claims. The scientific part of the development of Bt. cotton is well documented. But there could be high expectations emanating from the farmers and the society. It would therefore be very important to educate the farmers and the dealers in the right way so that they could aim for realistic expectations from Bt cotton. Farmers must be told that Bt cotton does not control all pests. It reduces the target pests substantially, but it does not eliminate or eradicate the sucking pests and all other pests. Consequently, there may be need for only one or two need based application of insecticides to control particularly sucking pests. While planting Bt. cotton, farmers must undertake extensive scouting of the field to decide and understand when supplemental sprays of chemical insecticides for control of cotton sucking pests are necessary. As this is a new technology for the country, such caution approach is called for to disseminate the technology in the Indian environment.

Furthermore, opposition to products of biotechnology stems from sheer ignorance or malice rooted in a hidden agenda. Many stakeholders are in the dark. It is high time that the scientists involved in the development of GM products and the biotech industry, which has been silent all the while, should come forward to promote awareness and educate the public and the farmers on the benefits of GM crops. The Central and State Governments should assume the role of leaders in creating public awareness about biotechnology.

BT COTTON CULTIVATION IN MANY COUNTRIES

GM cotton has become widespread, covering a total of 15 million hectares in 2007, or 43 percent of the world's cotton. Most GM cotton is grown in India and the US, but it can also be found in China, Argentina, South Africa, Australia, Mexico, and Columbia. The GM cultivars grown today are resistant to herbicides or insect pests.

More than half (68%) of China's cotton production is genetically modified to produce a substance (Bt toxin) that protects it against insect pests. A few types of caterpillars are especially problematic because they bore into cotton bolls reducing yield and compromising quality. Cotton used to be protected from insects by repeated pesticide applications. Bt cotton has now enabled Chinese farmers to dramatically reduce pesticide use.

The production of GM cotton has not yet been approved in the EU. Applications have been submitted, but a decision is still pending. Several lines of GM cotton have been approved in the EU, but only for use as food and feed.

EXPERIENCE IN OTHER PARTS OF THE WORLD

Bt cotton has already been commercialized in six countries: the United States (1996), Australia (1997), South Africa (1997), Argentina (1998), Mexico (1996), China (1998), and Indonesia (2000). Bt cotton is being extensively field tested under permit in Brazil, Colombia, Thailand, and Zambia. Globally, Bt cotton was planted on one million acres (405, 000 ha) in 2000. Due to its excellent performance, the benefits to the farmer of Bt cotton have been consistently superior for the six years of commercialization.

BT COTTON IN INDIA

Table. Timeline summary for regulatory processes leading to commercial release of Bt cotton in India.

Years	Studies undertaken	Government of India oversight committees
1995-1996	Application and permit for importation of Bt cotton seed containing the *Cry1Ac* gene	DBT
1996-2000	Greenhouse breeding for integration of the *Cry1Ac* gene into Indian germplasm, seed purification, and stock increase	DBT
1996-2000	Limited field studies for potential of pollen escape, aggressiveness, and persistence	RCGM (DBT)
1998-2001	Biochemical and toxicology studies	RCGM (DBT), GEAC
1998-2000	Multilocation field trials: agronomic and entomology performance of first-generation Bt cotton hybrids, conducted by Mahyco and State agriculture universities	RCGM (DBT), MEC
2000-2001	Soil rhizosphere evaluations and protein expression analyses from multilocation field trials	RCGM (DBT), GEAC
2001	Advanced stage multilocation field performance trials of first-generation Bt cotton hybrids, conducted by ICAR	GEAC, ICAR, DBT, MEC
2002	Submission of final biosafety, environmental safety, gene efficacy and performance documentation to GEAC; commercial release of first-generation Bt cotton hybrids by GEAC	GEAC
2002-ongoing	Continued field performance trials of second-generation Bt cotton hybrids for regulatory approval	RCGM (DBT), GEAC, ICAR, MEC

DBT = Department of Biotechnology; GEAC = Genetic Engineering Approval Committee; RCGM = Review Committee for Genetic Modification (constituted by DBT); ICAR = Indian Council of Agriculture Research; MEC = Monitoring & Evaluation Committees (constituted by GEAC and RCGM).

Bt cotton seeds were first tested in India for germination, vigor, and insect efficacy. Other experiments were conducted to confirm the environmental safety of Bt cotton, including tests of gene flow, persistence of the transformed plants, weediness characteristics, crossability of the transgenic pollen with the nontransgenic relative and near relatives, effect of the pollen on insects and nontarget organisms, and changes in the soil microbial flora. These studies were conducted under the unique environmental conditions of India and with the Bt trait in Indian germplasm. Studies of the molecular characterization and stability of the *Cry1Ac*gene were also carried out, as well as feeding studies and tests of food and feed safety, toxicity, and allergenicity.

BT COTTON—AGRONOMIC BENEFITS

Trials conducted in several locations in 1998/99, 1999/2000, 2000/01, and 2001/02 demonstrated the following agronomic benefits of Bt cotton:

- Good control of bollworm species in different growing areas;
- Significantly higher yield and boll retention (compared to control or non-Bt cotton);
- Reduction in expense of insecticide application;
- Additional revenue (Rs.2,500-4,000/acre) in farm income (compared to non-Bt cotton); and
- no adverse effect on nontarget insects or adjacent non-Bt cotton crops.

BT COTTON—APPROVAL FOR COMMERCIAL CULTIVATION

The Genetic Engineering Approval Committee (GEAC), in its 32meeting, held on March 26, 2002, made the landmark decision of approving cultivation of Bt cotton in India; three hybrids (MECH 162 Bt, MECH 184 Bt, and MECH 12 Bt) were approved for cultivation with the stipulation that certain conditions be met. These hybrids are high yielding and produce medium-long- to long-staple fiber.

Fig. Comparison of a Bt cotton hybrid (right) with its non-Bt hybrid counterpart (left) at first picking stage during a regulatory field trial in India. Three intra-*hirsutum* Bt cotton hybrids were approved for commercialization by the government of India in 2002.

The GEAC also approved production of seed; subsequently, seed for cultivating 100,000 acres (40,485 ha) was available for planting by Indian farmers. The seed was distributed in the states of Maharashtra, Madhya Pradesh, Karnataka, Andhra Pradesh, Gujarat, and Tamil Nadu in the kharif 2002 planting season. The areas planted to Bt cotton were 14746, 2220, 3908, 5608, 8854, and 2702 hectares spread over the three hybrids.

Table. Commercial cultivation of Bt cotton hybrids in India, 2002 (hectares).

State	MECH-12	MECH-162	MECH-184	Total
Maharashtra	112	9,300	5,334	14,746
Madhya Pradesh	60	404	1,756	2,220
Karnataka	—	3,828	80	3,908
Andhra Pradesh	44	5,564	—	5,608
Gujarat	76	4,136	4,642	8,854
Tamil Nadu	—	2,042	660	2,702
Total	292	25,274	12,472	38,038

This year's cotton-growing season in India was affected by high rainfall in some areas and a long dry spell followed by heavy downpours (resulting in unfavorable conditions for cotton cultivation) in other areas. The overall pest pressure this year for bollworm complex was also low. However, across the cotton-growing areas, control of the bollworm complex was observed in Bt cotton; the amount of spraying required, if any, was significantly lower for Bt than for non-Bt cotton hybrids. Bt cotton yields were significantly higher than those of non-Bt cotton, and the average increase in yield was about 30% over non-Bt hybrids in similar conditions. The quality of Bt cotton was cleaner with better color, and Bt cotton provided rates commensurate with the quality of the fiber based on the hybrids available.

The substantial increase in yield, there is a significant decrease in the number of insecticide sprays associated with the use of Bt cotton—the overall average indicates a yield increase of 8.1 quintals of cotton and a reduction of 1.93 sprays. These two factors add to the total economic benefit. An average additional income of more than Rs.18,000/ha for Bt compared to non-Bt cotton.

PROJECTION OF BT COTTON FOR NEXT THREE YEARS

Table. Projection of Bt cotton planting area in India (hectares).

Hybrids	2002/03		2003/04		2004/05	
	Area	%[a]	Area	%[a]	Area	%[a]
MECH-12	292	0.01	80,000	1.66	80,000	1.66
MECH-162	25,274	0.52	160,000	3.33	160,000	3.33
MECH-184	12,472	0.25	28,000	0.58	40,000	0.83
New hybrids[b]	—	—	40,000	0.83	280,000	5.83
Total	38,038	0.78	92,000	6.40	560,000	11.65

[a] % = percentage of area under hybrid cotton. Percentages are based on the present total of 4.8 million hectares of hybrid cotton area in India.

[b] New hybrids will be made available only on the approval of the Ministry of Environment and Forests.

Today, Bt cotton comprises 0.78% of the hybrid cotton area. It is projected that the 2003/04 and 2004/05 seasons will have Bt coverage of 6.40% and 11.65%, respectively.

CONCLUSION

The benefits of Bt cotton in India are in line with those enjoyed by farmers worldwide who have cultivated Bt cotton. The area under Bt cotton cultivation is expected to increase—it is likely that an area of 500,000 hectares will be covered by 2005, leading to increased production and reduced costs in an environmentally favorable manner. This will positively affect the livelihood of millions of small farmers by improving their net incomes.

Bt cotton is undoubtedly the most extensively studied cotton variety today. Rigorous scientific studies conducted in India and abroad demonstrate that Bt cotton and its products are safe for the environment, humans, animals, and agriculture. In fact, the use of Bt cotton is a positive step towards environmental protection because it makes possible the reduction of the insecticide load in the environment and reduces handling of such chemicals by farmers. This reduced use of insecticides will enhance the effectiveness of biological controls and implementation of Integrated Pest Management (IPM) programs. The higher farm income observed in the experiments has now been demonstrated by the large-scale use of Bt cotton by Indian farmers, and the incorporation of the gene is proving an effective and environmentally friendly plant protection tool resulting in greater cultivation of Bt cotton in the coming years. The cotton trade is looking forward to the productivity and quality benefits of Bt cottonseed. Efforts are being made to incorporate another gene (Bollgard II) to improve efficacy and postpone possible resistance problems. As newer products are approved in the regulatory system, it is likely that farmers will have greater choice to plant hybrids according to market quality requirements.

POOR CROP MANAGEMENT PLAGUES *BT* COTTON EXPERIMENT IN INDIA

Scientists in India are trying to figure out the ramifications of the growth of scores of illegal, untested, genetically modified varieties of cotton alongside legal varieties. Combined with a bad monsoon and the government's failure to educate farmers and regulate effectively, this is clouding an objective assessment of the first-year performance of GM cotton—a situation that could hinder future progress of the technology in India.

Farmers in five Indian states are cultivating Monsanto's (St. Louis, MO) *Bt* cotton on over 100,000 acres after India's Genetic Engineering Approval Committee (GEAC) gave the go-ahead in March to three hybrids developed by the company. The seeds carry the *Bacillus thuringiensis* (*Bt*) gene and produce a natural pesticide lethal to the bollworm, a scourge of cotton fields worldwide.

Farmers have bought *Bt* cottonseed at four times the price of traditional varieties in the hope it will bring them better returns.

Both the government and Mahyco-Monsanto Biotech India Ltd. (Mumbai)—the joint venture between Monsanto and Maharashtra Hybrid Company that sold the seeds—claim the crop is doing well. "The truth is that we have very positive feedback on Bollgard (*Bt*cotton) from farmers in all the cotton-growing states in the Centre and the South," Ranjana Smetacek, Monsanto's spokesperson, told *Nature Biotechnology*. "The agriculture ministry is happy, the evaluation committee is happy, and I can show you excellent photos [of the *Bt* fields] I have received," added Department of Biotechnology secretary Manju Sharma.

However, reports from non-government organizations (NGOs) suggest the crop is failing. Officials in Andhra Pradesh say that *Bt* cotton in the state is underperforming, and Gujarati newspapers have reported that there has been heavy bollworm infestation of *Bt*cotton, which was also found susceptible to leaf-curl virus and root-rot disease, and that in Madhya Pradesh, *Bt* cotton suffered greater damage due to drought than traditional varieties grown there.

One of the problems, according to both government sources and NGOs, is that local farmers are not meeting the many technical specifications—such as for refugia management and planting conditions—for *Bt* cotton, a relatively high-maintenance crop. Cotton farmers with very small land holdings, for instance, have found it impossible to set aside land for refugia, and only 40% of the total area of cotton is irrigated—which is causing problems this year because of a delayed monsoon.

Prasantha Kumar Ghosh, a former advisor in the Department of Biotechnology, says *Bt* cotton is facing problems this season because of poor rains. "*Bt* cotton is input intensive, and our trials have clearly shown this."

Suman Sahai, convener of Gene Campaign, a Delhi-based NGO, and a visiting professor at the University of Heidelberg, blames the government and scientific community for failing to educate farmers about dangers of not following proper procedure. "We have consistently argued that any new technology must be introduced only after farmers and consumers have complete information on all its aspects so that they can make an informed choice." Sahai says the government has still not placed in the public domain data generated by trials of *Bt* cotton in India. Devinder Sharma, anti-GM campaigner and director of the Forum for Biotechnology and Food Security (FBFS) in New Delhi, agrees. "The GEAC is solely responsible for hastily pushing the untested technology," says Sharma.

To make matters worse, several thousand acres—even in areas such as Punjab and Haryana where *Bt* cotton has yet to be approved—have been sown with second and third generations of a Monsanto knockoff known as Navbharat

151, according to Sahai. Last year, these seeds were covertly sold by the Navbharat Seeds Company and planted by Gujarati farmers on over 10,000 acres (*Nat. Biotechnol.* 19, 1090, 2001). Illegal *Bt* seeds (that the government failed to destroy) from last year's harvest in Gujarat have been flooding the market at one-tenth to one-half the price of legal seeds. First-generation seeds do not have the same vigor as the originals, and subsequent generations have even worse quality and yield.

The presence of so many varieties of *Bt* cotton "is making a public mockery of India's ability to regulate and direct the use of this new and controversial technology," says Sahai. "The government must recognize the chaos it has created and take corrective steps."

"The large quantity of untested and unauthorized *Bt* hybrid seeds are likely to cause losses, and farmers are likely to lose faith in *Bt* cotton, which would damage this useful technology," says Arvind Kapur, managing director of Numhens Proagro, which is getting ready to seek government approval for its GM mustard. However, A.S.N. Reddy, a senior official at Proagro, believes there is no need to panic. "In about two years the situation will stabilize," he said. "Once farmers realize the fake *Bt*-cotton varieties they use are no good, they will turn to the genuine ones."

Meanwhile, Mihir Shah, director of the Baba Amte Centre for People's Empowerment, and Debashish Banerji of the Samaj Pragati Sahyog (Nature and Society Cooperative), based in Madhya Pradesh, say: "This is obviously not a technology meant for the poor, dryland small farmers of India." Whether or not this is true will be determined from a full analysis by the Ministry of Agriculture at the end of the year, after the main harvesting season, which begins this month.

BT-*COTTON IN INDIA: THE TECHNOLOGY WINS AS THE CONTROVERSY WANES*

Cotton is an important cash crop in India and plays a significant role in the national economy. As an industrial crop, it supports millions of people through cultivation, processing and trade and contributes Rs.360 billion (US$8 billion) to the export income. The area occupied by cotton in recent years fluctuated between 8 to 9 million hectares (1 hectare = 2.471 acres) in India. While India has the largest area under cotton in the world (representing 20 to 25% of the global area), it ranks only third in terms of production after China and USA. Several factors are responsible for such low yields of which losses due to insect pests are the most important. More than 160 species of insects have been reported to attack cotton at various stages of its growth as defoliators, tissue borers and sap-suckers, causing losses up to 60%. Among them, bollworms (tissue borers) are the most destructive, requiring major efforts to save the crop from them.

COTTON BOLLWORMS

The cotton bollworm complex in India includes the 'old world bollworm' or 'false American bollworm' - *Helicoverpa armigera*; pink bollworm - *Pectinophora gossypiella*; spotted bollworm -*Earias vittella* and spiny bollworm - *Earias insulana* . All these belong to the insect order

Lepidoptera. The tobacco caterpillar - *Spodoptera litura*, also a lepidopteron, is a sporadic pest on cotton. Although predominantly a defoliator, it can also damage cotton bolls and squares when there is a severe outbreak.

Among the bollworms, *H. armigera* is the most dominant and difficult to control chiefly due to its widespread insecticide resistance, multivoltine and prolific pattern of breeding and high polyphagy. It is a highly destructive and wasteful feeder in the sense that a single larva can damage many squares and bolls. *H. armigera* has a wide distribution, but is limited to the old world i.e., Europe, Asia, Russia, Africa, Australasia and the Pacific Islands. This species does not occur in the Americas. The species found in the Americas are *Helicoverpa zea* and *Heliothis virescens,* popularly called 'bollworm' and 'tobacco budworm,' respectively. Hence, reference to *H. armigera* as 'American bollworm' is misleading. To avoid any confusion, it is better to call *H. armigera* as 'old world bollworm' or 'false American bollworm.'

Chemical insecticides are used extensively on cotton crop for control of insect pests, especially bollworms. The number of sprays per crop season may vary from 5 to 20 or more. It is estimated that insecticides worth about Rs.30 billion (US $ 660 million) are used annually in Indian agriculture of which about Rs.16 billion are spent for the control of cotton pests and of this Rs.12 billion against bollworms alone. In terms of volume, about 54% of the total insecticides used in Indian agriculture are sprayed on cotton crop. This indicates the economic importance of bollworms in general and *H. armigera* in particular. Despite such huge efforts, bollworm control has not been generally satisfactory mainly because a pest like *H. armigera* has developed resistance to most of the currently recommended insecticides. Nevertheless, farmers continue to use insecticides repeatedly as they have no option except to *'Spray' or 'Pray.'* This had frustrated the farmers, scientists and policy makers alike. *Bt*-cotton came at a time when they were desperately looking for an alternative and dependable control measure.

DEVELOPMENT OF *BT*-COTTON (BOLLGARD®) IN INDIA

Realizing the economic importance of cotton bollworms and the benefits *Bt*-cotton can offer to the growers, MAHYCO (Maharashtra Hybrid Seed Company), a leading Indian seed company, took the initiative in introducing this technology into India in collaboration with Monsanto Company.

As per regulatory procedure, Mahyco sent its application to Dept of Biotechnology (DBT), Government of India, in March 1995 seeking permission for introducing this technology. On obtaining approval, Mahyco received 100gms of *Bt*-cotton seeds (variety Cocker 312) containing the Bollgard® *Bt* gene, *cry 1Ac,* from Monsanto, USA, in March 1996. These seeds were first tested in India under greenhouses for germination, vigour of the plants and their efficacy against the Indian cotton bollworms. These were used in the greenhouse breeding programmes and 40 elite Indian parental lines were introgressed with *cry 1Ac* gene by crossing with the *Bt* gene donor parent obtained from Monsanto. Mahyco had already developed several cotton hybrids suitable for different agro-climatic regions and these were popular with the farmers. Some of these ruling conventional hybrids were converted into Bollgard® (this is the brand name of *Bt*-cotton developed by Monsanto) using the converted parental lines and tested for their performance and safety.

REGULATORY STUDIES ON SAFETY

In India, two union ministries are involved in the regulation of GMOs – Ministry of Science & Technology and Ministry of Environment & Forests. The Department of Biotechnology (DBT) functions under MoST. Two important committees, namely Institutional Bio-Safety Committee (IBSC) and Review Committee on Genetic Modification (RCGM), work under the guidance of DBT. Another major committee, namely Genetic Engineering Approval Committee (GEAC), is constituted under MoEF. These committees are represented by experts drawn from various fields and organizations across the country and are responsible to ensure that proactive safety studies are carried out on GM products before they are approved for commercialization.

As per the direction and guidelines of the regulatory authorities, a number of studies were carried out to assess the safety of *Bt* protein expressed in *Bt*-cotton (Bollgard®) plants with regard to its potential for allergenicity, toxicity, gene flow, cross pollination, effect on non-target beneficial organism, impact on soil microorganisms etc. These data were examined by the expert committees.

Feed-safety studies with *Bt* cottonseed meal were carried out with goats, buffalos, cows, rabbits, birds and fish. The results revealed that the animals fed with *Bt*-cotton seed meal were comparable to the control animals in various tests and showed no ill-effects. These studies were carried out by the Industrial Toxicological Research Centre, Lucknow; National Dairy Research Institute, Karnal; Central Institute of Fisheries Education, Mumbai; Central Avian Research Institute, Bareily; National Institute of Nutrition, Hyderabad; and Govind Vallabh Pant University for Agriculture and Technology, Pantnagar.

Studies were also conducted on the effect of leachate from *Bt*-cotton plant on soil rhizosphere and non-rhizosphere microflora, soil collembola and earthworms. The results showed no difference between the soils where *Bt* and non-*Bt* plants had been grown. The information generated on pollen dispersal has established that airborne pollen transmission is limited to only a couple of meters and the risk of undesirable introgressive hybridization with related species is minimal. Further, *Bt*-cotton hybrids are tetraploid in genetic composition whereas their nearest relatives, the local '*Desi*' cotton varieties, are diploid and hence genetically incompatible for hybridization. Studies conducted have also revealed that *Bt*-cotton had no adverse impact on biological control agents like ladybird beetles, green lacewings and parasitic hymenoptera.

Studies were also carried out to determine the levels of *Bt* protein expressed in different tissues (terminal leaves, squares and bolls) at different age of the crop and at different locations. The results revealed that although the expression varied between the tissues and also with the age of the plant, the amount of protein present in various tissues at any given time was adequate to bring about mortality of the early instar bollworms.

Baseline susceptibility data were also generated for a number of geographic populations of *Helicoverpa armigera* so that it can serve as a benchmark for monitoring resistance, if any, in future. These studies were carried out, prior to commercial cultivation of *Bt* cotton, by Project Directorate of Biological Control, ICAR, Bangalore. Field trials conducted from 1998 to 2001 have clearly indicated that *Bt*-cotton hybrids provided effective control of the bollworm complex in all the locations and seasons. Data generated on all these aspects were submitted to DBT / RCGM for review.

INDIA APPROVES *BT*-COTTON – THE FIRST AGRIBIOTECH PRODUCT

Based on the recommendation of RCGM, the Genetic Engineering Approval Committee (GEAC), in its 32nd meeting held in New Delhi on 26th March 2002, approved Mahyco's *Bt*-cotton for commercial cultivation, pronouncing it to be safe and beneficial. This is a landmark decision as *Bt*-cotton happens to be the first-ever agricultural biotech product to receive official approval and with it India made its long awaited entry into commercial agricultural biotechnology. This approval has specified three *Bt* hybrids, namely Mech 12, Mech 162 and Mech 184 which had undergone all the trials and it was initially granted for three years. The approval also stipulated certain other conditions and one of them was that every *Bt*-cotton field shall be fully surrounded by a 'refuge' crop comprising the same non-*Bt*-cotton hybrids and the size of the refuge shall be at least five rows of non-*Bt* or 20% of the total sown area whichever is greater. The idea of 'refuge' is to serve as a strategy to produce *Bt* sensitive insects thereby helping to prevent or delay the development of resistance by bollworms to the *in planta* produced *Bt* protein. Besides, 'refuge' can act as a 'pollen sink' area to some extent.

The chronology of events that led to the development and approval of *Bt*-cotton in India are summarized below.

Table. Chronology of Development and Approval of *Bt*-Cotton in India

1995	Mahyco applied to DBT (Department of Biotechnology, Govt. of India) for permission to import a small stock of Bollgard® (*Bt* cotton) seeds from Monsanto Company, USA. DBT gave permission.
1996	A nucleus stock of 100 gms of cotton seeds of the variety Cocker 312 containing the Bollgard® *Bt* gene, *cry 1Ac*, was received by Mahyco from Monsanto, USA. Initiated crossing with the Indian cotton breeding lines to introgress *cry 1Ac* gene. 40 elite Indian parental lines were converted for *Bt* trait.
1996-1998	Risk-Assessment Studies conducted using *Bt*-cotton seeds from converted Indian lines. - Pollen escape studies - Aggressiveness and persistence studies - Biochemical analysis - Toxicological studies on ruminants (goats) - Allergenicity study on rabbits
1998 – 1999	Multi-location field trials at 40 locations in 9 states to assess agronomic benefits and safety. Data submitted to RCGM (Review Committee for Genetic Modification), Ministry of Science & Technology, Govt. of India.
1999 – 2000	Field trials repeated at 10 locations in 6 states. Data submitted to RCGM.
2000	July 2000 – Based on the recommendation of RCGM, the GEAC (Genetic Engineering Approval Committee), Ministry of Environment & Forests, Govt. of India, gave approval for Mahyco to conduct large scale field trials in 85 ha and also undertake seed production in 150 ha.
2001	*Kharif* 2001 – Large scale field trials covering 100 ha. Field trials were also conducted by All India Coordinated Cotton Improvement Project of the Indian Council of Agricultural Research (ICAR).
2002	On 26 March 2002, GEAC approved Mahyco's three *Bt*-cotton hybrids, viz. MECH 12, MECH 162 and MECH 184, for commercial cultivation in India. This approval was initially valid for three years and also stipulated a few conditions. **This is a landmark decision as *Bt*-cotton is the first-ever transgenic crop to receive such a regulatory approval in India.**

FIELD PERFORMANCE

In the year 2002, the first year of approval, three *Bt*-cotton hybrids, Mech 12, Mech 162 and Mech 184, were commercially planted on about 29,415 ha (72,685 acres) in six states - Maharashtra, Madhya Pradesh, Karnataka, Andhra Pradesh, Gujarat and Tamil Nadu. The area increased to 86,240 ha in 2003 and to 530,800 ha in 2004. The results demonstrated the following benefits from *Bt*-cotton:

- Good control of bollworm species (false American bollworm, pink bollworm, spotted bollworm, spiny bollworm) in all locations and seasons
- Significantly higher boll retention and more yield than the control or non-*Bt* cotton crop

- Reduction in chemical sprays for bollworm control
- Substantial increase in net income to farmers
- No adverse impact on non-target organisms and the adjacent non-*Bt* cotton or other crops

During the growing season of *Kharif* 2003, commercial performance trends of *Bt*-cotton were tracked by Mahyco for approximately 3,000 farmers covering most cotton growing states in central and south India. Data were taken for all three released Bollgard® hybrids, namely Mech 12, Mech 162 and Mech 184. The largest sample was taken in the state of Maharashtra due to greater availability of resources.

From a total sample size of 1,700 Maharashtra farmers, trends for relative economic gain in favour of Bollgard® hybrids ranged from Rs.15,854 to 20,196 (av. Rs.18,325) per hectare among the three hybrids. For all Bollgard® hybrids, the average number of insecticide applications for bollworm complex was about 50% less than that required for conventional commercial hybrids. Seed cotton yield in Bollgard® hybrids ranged from 19.00 to 21.7 (av.20.13) quintals (1 quintal = 100 kg) per hectare as compared to conventional hybrids where it varied from 11.00 to 13.09 (av. 12.23) quintals per hectare. Similar trends were documented in all the other states surveyed. The average net economic benefit from Bollgard® hybrids over non-*Bt* hybrids among all states in the survey ranged from Rs.9,995 to 32,889 per hectare.

A nationwide survey carried out by ACNeilsen-ORG MARG in 2003 which included 3,063 farmers (1672 *Bt* farmers and 1391 conventional farmers) from Maharashtra, Madhya Pradesh, Andhra Pradesh, Karnataka and Gujarat (Tamil Nadu could not be included as the harvest was yet to be completed) clearly indicated the benefits of Bollgar® cotton. It indicated that the *Bt*-cotton growers in India were able to obtain, on an average, a yield increase by about 29% (range 18 to 40%) due to effective control of bollworms, a reduction in chemical sprays by 60% (range 51 to 71%) and an increase in net profit by 78% (range 66 to 164%) as compared to non-*Bt* cotton. The net profit translates to an average of Rs.7,724 (range Rs. 5,900 to 12,696) per hectare. According to the survey, over 90% of Bollgard® users and over 40% of non-users expressed the intent to purchase Bollgard® seeds in the coming season. Similar trends and benefits had been reported from other countries also.

OPPOSITION TO *BT*-COTTON

Bt-cotton had faced opposition from a couple of organizations and a few individuals right from the beginning of its introduction and even before it could complete any regulatory studies. The issues raised were mostly speculative, complex and confusing and ranged from scientific - social, ethical - emotional, economical - egoistic, political - publicity, ignorance - arrogance, legal - illegal and reasonable - unreasonable!

A farmers' organization in Karnataka, namely Karnataka Rajya Raitha Sangha (KRRS), uprooted and burnt a few approved experimental crops in 1998 and also in 1999, wrongly accusing that *Bt*-cotton contained the so-called 'Terminator Technology' and the gene would escape and cause 'Gene pollution' and sterility in other plants!! They also alleged that *Bt* protein is harmful to humans, farm animals, other beneficial organisms and soil! Obviously they were oblivious of the real technology and the highly selective action of the concerned *Bt*-protein. They threatened the farmers with serious consequences if they planted *Bt*-cotton. They also held repeated public demonstrations against this technology.

There were also a few other critics who were busy projecting themselves as saviours of the environment and farmers. Whenever the cotton crop failed in certain areas, be it due to drought or other environmental stress, wilt or other diseases, sucking pests or any other reason, they mischievously attributed it to failure of the *Bt*-technology and blamed the company as well as the government. They tried to instigate the farmers to claim compensation from the company, ignoring the fact that *Bt*-cotton has been developed specifically to offer protection against bollworms, not against any other adverse factors. Their action and statements had received prominent coverage in the print and electronic media and created a lot of doubt and confusion in the minds of innocent farmers and public. *Incorrect knowledge is more dangerous than ignorance!* It took enormous efforts on the part of Monsanto and Mahyco to mitigate such negative publicity and make people realize the true value of *Bt*-cotton. The role played by DBT, which stood by this technology and organized several educational seminars on biotechnology in several states in this regard, is commendable. Except for a very few scientists, the rest of the scientific community remained silent when this emerging technology was so unreasonably attacked and misinformation was spread!!

The practical results obtained in India with *Bt*-cotton demonstrated that it is safe and beneficial and were comparable with those in other countries where it had been commercialized earlier, starting from 1996 in the USA. Several scientific publications are available in this regard. Yet, the opponents are continuing their tirade against this technology and churn out their own data. They seem to thrive on controversies and try to be in the news somehow. *Protesting has become a profession!* However, of late, it has little impact on the farmers as they have personally cultivated or observed *Bt*-cotton and realized the benefits. *It is apparent that as the technology wins, the controversy wanes.*

ILLEGAL *BT*-COTTON IN INDIA

Realizing the potential of *Bt*-cotton in India, certain unscrupulous agencies are exploiting the situation through sales of unapproved *Bt*-cotton or spurious seeds. In fact, such seeds were introduced into the market while Mahyco was still carrying out the regulatory trials and waiting patiently for government approval. It was first discovered in Gujarat in 2000 and Navbharat Seeds Pvt. Limited was identified as the offending company. Later, such illegal seeds were found in several other states also where they occupied, and continue to occupy, several thousand hectares. It has the following implications:

- Unapproved research on GMOs and commercialization is a blatant violation of bio-safety norms and is a punishable offense
- Spurious producers are not accountable for purity, performance and safety. They may spoil the credibility of the product and technology
- Can afford to sell their products at a much lower price as their investment on research is
- meagre
- Will affect the confidence and enthusiasm of genuine technology developers who invest a lot of time, talent and money in developing new products and getting their approval through due regulatory procedures
- Users will be misled and confused

Illegal *Bt*-cotton is a blatant contravention of bio-safety norms and business ethics. Although the government has shown some concern and initiated action, this serious issue needs to be curbed more urgently and more strictly with severe penalty.

PROSPECTS FOR *BT*-COTTON

Bt-cotton was first commercialized in the USA in 1996 and subsequently in other countries like Australia (1996), Argentina (1997), China (1997), Mexico (1998), South Africa (1998), Colombia

(2002) and India (2002). As of 2003, it was cultivated in 7.2 million hectares in eight countries. Substantial increase in yield due to effective control of bollworms, considerable reduction in chemical sprays and significant increase in net profit to farmers are a common feature in all the countries. The area under *Bt*-cotton will increase considerably in all the countries in the coming years.

In India, bollworms are a major threat to cotton crop. Hybrid cotton crop is more severely attacked than the local varieties. The total area under cotton in India is about 9.0 m ha of which 4.8 m ha are occupied by hybrids and the rest by the varieties. In the year 2004, *Bt*-cotton occupied only 0.53 m ha which constituted less than 6% of the total cotton area or 11% of the hybrids. The indications are that the area will continue to increase significantly in the coming years. Realizing the potential, about 19 other seed companies in India, who have their own cotton hybrids suited for different regions, have already joined Mahyco & Monsanto as their sub-licensees for *Bt*-cotton. Their hybrids as well as Mahyco's new hybrids incorporated with *cry1Ac* are already undergoing regulatory trials. In fact, a *Bt* hybrid, RCH 2, developed by Rasi Seed Company has already received regulatory approval in 2004.

Mahyco is also carrying out regulatory trials with Bollgard II® stacked with two *Bt* genes, *cry 1Ac* along with *cry 2Ab2*, also licensed from Monsanto. Bollgard® II has already received commercial approval in Australia in September 2002 and in the USA in December 2002. It is superior to Bollgard® in performance and host range (in addition to other bollworms, it is also effective against *Spodoptera* spp.) and also makes a very good product for insect resistance management (IRM).

Planting refuge is mandatory in India as in the USA, Australia and other countries as a strategy towards insect resistance management. In India, *Helicoverpa armigera*, by far the most predominant bollworm attacking cotton, also infests a large number of other crops like chickpea, pigeonpea, tomato, sunflower, maize and sorghum. These crops occupy substantial area and are cultivated around the cotton crop at the same time in several parts of south and central India. These crops, especially chickpea and pigeonpea, support larger populations of *H. armigera* than cotton, thereby serving as natural refuge and helping IRM. Further, as the area presently occupied by *Bt*-cotton is very small (i.e., less than 6% of the total cotton area or 11% of hybrid cotton), a huge crop of non-*Bt* hybrids and varieties are also available as refuge. In view of this, it appears that growing non-*Bt* cotton as structured refuge may not be required in India. In fact, in China, for the same reasons, structured refuge is not mandatory.

Bt-cotton is a well-researched scientific product. The facts reveal that in the last 7-8 years of its commercial cultivation in various countries, it has brought significant economic and environmental benefits and did not cause any untoward incidents related to bio-safety, environment or pest resistance. Despite, negative reports continue to be hoisted by certain individuals. While it may not be worthwhile trying to convince such blind opposition (*"You can wake up a sleeping person, but cannot wake up a person who is pretending to be asleep"*), efforts should be made to prevent their misleading influence on innocent farmers and other people. This will continue to be a tough challenge for biotechnology and call for greater efforts towards biotech awareness and education. Scientific outreach is a highly skilled job where *science should be simplified while communicating with common people*. Healthy criticism is welcome and we should always strive for improvement. *Bt*-cotton is a remarkable product and our farmers should be encouraged to derive maximum benefit from it like thousands of farmers in other countries.

GLOBAL IMPACT OF INSECT-RESISTANT (BT) COTTON

Insect-resistant cotton was first introduced commercially in 1996. It is commonly referred to as Bt cotton, because it produces an insecticidal protein from the naturally occurring soil bacterium *Bacillus thuringiensis* (Bt). Global adoption of Bt cotton has risen dramatically from 800,000 hectares in its year of introduction in 1996 to 5.7 million hectares (alone and stacked with herbicide-tolerant cotton) in 2003. Significant economic and production advantages have resulted from growing Bt cotton globally. Bt cotton can substantially reduce the number of pesticide sprayings, which reduces worker and environmental exposure to chemical insecticides and reduces energy use. The quality of life for farmers and their families can be improved by the increased income and time savings offered by Bt cotton. These economic, environmental, and social benefits are being realized by large and smallholder farmers alike in eight countries around the world.

DEVELOPMENT OF BT COTTON

Bt cotton produces an insecticidal protein (*Cry1Ac*) from the naturally occurring soil bacterium *Bacillus thuringiensis* (Bt) that protects the cotton plant from certain lepidopteran (caterpillar) insect pests (Perlak et al., 2001). Coker 312 cotton was transformed to express the *Cry1Ac* gene from Bt, resulting in cotton plants that were resistant to attack from major lepidopteran pests (Perlak et al., 1990). Many years of development followed to deliver the trait in germplasm varieties that meet the strict agronomic requirements of growers worldwide (Perlak et al., 2001). In the United States, Monsanto's Bt cotton is known as Bollgard® cotton.

Extensive testing of Bt plants has demonstrated their safety and advantages (Betz, Hammond, & Fuchs, 2000). The food, feed, and environmental safety of Bollgard®cotton was evaluated by regulatory agencies prior to commercialization. Regulatory approval has been granted in countries where Bt cotton is grown as well as in countries that import Bt cottonseed products. Studies were conducted on the safety of the produced proteins, food/feed composition, and environmental safety. On the basis of this evaluation, Bollgard® cotton and its processed fractions were found to be substantially equivalent to conventionally bred cotton, and the Bt protein was shown to be safe for human and animal consumption. Bt cotton was found to pose comparable or fewer risks to the environment than traditional cotton treated with commercially approved insecticides. Safety data on Bollgard® has been provided to additional regulatory authorities globally, and regulatory review continues in these countries.

ADOPTION OF BT COTTON

Global adoption of Bt cotton has risen dramatically from 800,000 hectares in 1996 to 5.7 million hectares (alone and stacked with herbicide-tolerant cotton) in 2003 (James, 2003). In 2002, Bt cotton was grown commercially in the United States, Mexico, Argentina, South Africa, China, India, Australia, and Indonesia, and precommercial plantings were grown in Colombia (James, 2002). Bt cotton is a global product, with plantings in North America (United States), Australia, three countries in Latin America, one country in Africa, and three countries in Asia. Large-acreage farmers in industrialized countries (such as the United States and Australia) derive significant value from Bt cotton, but the vast majority of growers are in developing countries. Over six million Bt cotton farmers are in developing countries; the vast majority of these are resource-poor farmers in China and South Africa (James, 2002). A number of studies have examined the significant economic, environmental, and social benefits derived from growing Bt cotton (International Service for the Acquisition of Agri-biotech Applications, 2002; Ismael, Bennett, & Morse, 2002a, 2002b; James, 2002; Pray, Huang, Hu, & Rozelle, 2002; Purcell, Oppenhuizen, Wofford, Reed, & Perlak, 2004).

ECONOMIC AND PRODUCTION BENEFITS

Significant economic advantages have resulted from growing Bt cotton around the world. Bt cotton provided US farmers with an average net income increase of $20/acre and increased the total net value of US cotton production by $103 million in 2001 (Gianessi, Silvers, Sankula, & Carpenter, 2002). In China, net revenue increases have ranged from $357/hectare to $549/hectare in the three years studied when one compares Bt cotton with non-Bt cotton (Pray et al., 2002). In South Africa, smallholder farmers in the Makhathini region raised their yields and reduced their application costs, netting an economic advantage for Bt cotton growers of about $25-51/hectare (Ismael et al., 2002a, 2002b). Yield advantages have been noted in a number of studies, ranging from 5-10% in China, more than 10% in the United States, and more than 20% in four other countries (James, 2002). A recent report found that in field trials in India, average yields for Bt cotton hybrids were 80% greater than non-Bt hybrids (Qaim & Zilberman, 2003), although other results from India are less dramatic (James, 2002). Production advantages can result from the level of insect control achieved and the time savings and reduced labor needs that may result (Benedict & Altman, 2001; Edge, Benedict, Carroll, & Reding, 2001). Bt cotton is a valuable option for growers, as it provides superior pest control of pests with several features that provide advantages over other insect control agents (Benedict & Altman, 2001; Edge et al., 2001; Perlak et al., 2001).

ENVIRONMENTAL BENEFITS

Bt cotton can substantially reduce the number of pesticide sprayings, which can provide significant environmental benefits. A number of studies have demonstrated that insecticide sprays are reduced by using Bt cotton (Carpenter et al., 2002; Edge et al., 2001; James, 2002). Growers in the United States reduced insecticide use by 1,870,000 pounds of active ingredient (AI) per year in 2001 (Gianessi et al., 2002). In China, insecticide applications were reduced by an average of 67% and the kilograms of active ingredient by 80% (Huang, Rozelle, Pray, & Wang, 2002), while South African growers reduced sprays by 66% (Ismael et al., 2002a). The use of Bt cotton in place of conventional systems can positively impact nontarget organisms (NTOs) and beneficial organisms by preserving populations (Head et al., 2001; Smith, 1997; Xia, Cui, Ma, Dong, & Cui, 1999) and is compatible with integrated pest management initiatives (Benedict & Altman, 2001). In addition, Bt cotton adoption can provide secondary positive environmental impacts such as (a) saving on raw materials needed to manufacture chemical insecticides; (b) conserving fuel oil required to manufacture, distribute, and apply such insecticides; and (c) eliminating the need to use and dispose of insecticide containers (Leonard & Smith, 2001).

BENEFITS FOR SMALLHOLDER FARMERS

Bt cotton and other tools which lead to more productive agricultural systems can benefit smallholder farmers and their broader agricultural communities. At the macroeconomic level, the increased productivity can stabilize production and reduce risks for lenders. At the farm level, improvements in the insect control system being used can positively impact the quality of life for farmers and their families by increasing incomes, reducing insecticide sprayings, and offering savings in time (Ismael et al., 2002a; Pray et al., 2002). The nutritional demands of families may also be better met, as these families now have increased income that could potentially be used for more food purchases and food consumption. Time savings may be particularly important for women in South Africa, where women serve as heads of many of the households. The time saved by using Bt cotton may allow these women to care for children, elderly, or the sick or to engage in income-generating activities (Ismael et al., 2002a). Children were also a beneficiary of this technology, as those children in South Africa who no longer have to spray insecticides could now potentially devote more time to educational or other worthwhile pursuits (Ismael, 2002a). Water savings from using Bt cotton by reducing insecticide sprays are another source of significant benefits for smallholder farmers. The use of Bt cotton on a typical 1.7 hectare farm in the Makhathini Flats region of South Africa would result in a labor reduction of 12 days of spraying, eliminate 100 km of walking, and save 1,000 liters of water while increasing income by $85 (James, 2002). These cumulative benefits have dramatic social relevance for a segment of society that can benefit most.

TRANSGENIC COTTON IN MEXICO

In 1996, Mexico and the United States became the first two countries to plant Bt cotton commercially. Bt cotton area in Mexico reached 26,300 hectares (one third of the country's cotton area) during the 2000 growing season, with adoption varying from less than 10% in Sinaloa and Baja California to 96% in Comarca Lagunera. The same two Bt cotton varieties (NuCOTN 33B and NuCOTN 35B) that were introduced in the United States in 1996 through a strategic alliance between Monsanto and Delta and Pineland Co. (D&PL) have subsequently been marketed in five other countries, including Mexico.

In this paper, we summarize the impact of the introduction of Bt cotton in the Comarca Lagunera region in the northern states of Coahuila and Durango. The cotton area in Comarca Lagunera peaked at 142,777 hectares in 1944 but fell to less than 1,000 ha in 1992 and 1993 because of price and exchange rate volatility, changes in government policy, and a scarcity of water for irrigation. In 1994, Mexico's federal government and the state of Coahuila created a fund to reactivate the cultivation of cotton in the region. This fund provided for subsidized credit to producers through producer associations. At present, cotton yields in the Comarca Lagunera stand at 125% of the national average, having increased from less than one ton/ha in the late 1980s to 1.6 tons/ha in 2000. Yields had previously peaked in 1984, then declined due to serious problems with pest control. Only 8,283 ha of cotton were planted in the Comarca in 2000.

INSECT COMPLEXES, ADOPTION OF BT COTTON, AND CHEMICAL PESTICIDE USE

Seven important insect pests plague cotton in Mexico. The most damaging are pink bollworm (*Pectinophora gossypiella*), boll weevil (*Anthonomus grandis*), tobacco budworm (*Heliothis virescens*), and cotton bollworm (*Helicoverpa zea*), but fall armyworm (*Spodoptera exigua*), white fly (*Bemisia argentifolii*), and conchuela (*Chlorochroa ligata*) also cause crop damage and require treatment in some areas. Patterns of infestation levels and economic losses vary widely across the main growing regions and have been important determinants of the adoption of Bt cotton. Bt cotton is 100% effective in controlling two major pests—pink bollworm and cotton bollworm—and is partially effective in controlling tobacco budworm and fall armyworm. These four pests are often referred to as the budworm-bollworm complex (BBWC). Although annual infestation levels are variable, the BBWC are most damaging in Comarca Lagunera and Tamaulipas.

Pest populations vary from year to year as a result of weather conditions, cultural practices, and cropping patterns. Each year the government Plant Health Authority locates several dozen insect traps around Comarca Lagunera to monitor pest pressure. Pest infestation levels, particularly of boll weevil and pink bollworm, have fallen during the 1990s. Neither the pink bollworm nor the boll weevil has important plant hosts other than cotton, so effective cotton residue management and the high adoption rate of Bt cotton have been major factors in reducing pest populations in Comarca Lagunera. The near elimination of cotton during the 1992 and 1993 growing seasons has nearly eliminated the boll weevil. The government has provided financial support averaging $24/ha from 1998 through 2000 for pest control programs through the Regional Plant Health Committee. The focus has varied through the years, but programs have been carried out in extension, field pest pressure monitoring, post harvest control of cotton residue, and for subsidizing the adoption of Bt cotton.

The combined effect of the disappearance of the boll weevil, use of Bt cotton, and the reduced cotton acreage has been a dramatic fall in the use of chemical pesticides in Comarca Lagunera. The total amount of active ingredient (AI) applied to cotton in 1999 was just 2% of the amount applied in 1988, falling from 670,709 kgs to 11,842 kgs. Per-hectare pesticide use has fallen by more than 80%, from an average of nearly 14 kgs/ha of active ingredient in the 1980s to about two kgs/ha. The average number of pesticide applications for all insects has also fallen steadily, led by the decline in applications to control BBWC. Pesticide use is lower on Bt than conventional cotton varieties, but it seems clear that all cotton is under less pressure from BBWC than in the past, perhaps because of the widespread adoption of Bt cotton. This suggests that a new low-infestation level pest dynamic may be emerging in the region. Producers are still adjusting to a new approach to pesticide use, in which they are becoming increasingly reluctant to use chemical pesticides for fear of upsetting the new equilibrium between beneficial and destructive insects.

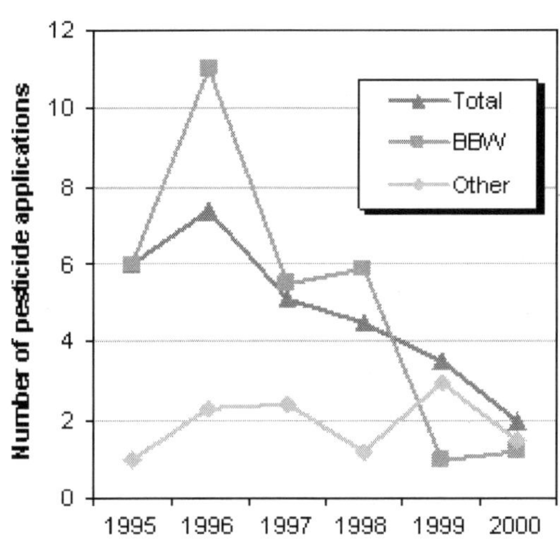

Note. Data from Sánchez-Arellano, 2000.

FINANCIAL BENEFITS OF BT COTTON IN COMARCA LAGUNERA

Bt cotton benefits in Mexico were modeled as occurring in a small open economy (Alston, Norton, & Pardey, 1995; Falck-Zepeda, Traxler, & Nelson, 2000). As holder of a patent on the technology, Monsanto/D&PL has a monopoly on the sale of Bt cotton, giving the firm the power to set seed prices above its marginal cost of production. Therefore, the welfare calculations performed below have two components—changes in farmer surplus and monopoly profits. Monopoly profit was calculated as Q_{Bt} (P_{Bt}-*c*), where Q_{Bt} and P_{Bt} are the quantity and price of Bt seed and *c* is the marginal cost of producing seed. We assumed that the market for conventional seed cotton is competitive, so that the market price represents the marginal seed production cost, *c*. Because no administrative, marketing, research and development, or intellectual property rights enforcement costs were deducted, these figures do not represent true surplus estimates, but rather represent gross Bt revenue.

BT AND CONVENTIONAL COTTON COST AND REVENUE DIFFERENCES

Producers in the region are generally classified as falling into two groups: ejidos and small landholders. The ejido producers (or *ejidatarios*) are very small producers whose holding was formed during one of Mexico's several land reforms. The size of the average ejido holding is 2-10 ha and that of the small landholders is 30-120 ha. The ejidos and small landholders are organized into farmer associations for the purpose of obtaining credit and technical assistance. The associations have centralized accounting, management, and technical staff. Each association comprises a number of smaller groups that farm together. Each farmer group is assigned a technical consultant, who makes most of the production decisions for the fields of all members of the group. In most cases, the individual landholders have relatively little involvement with actual production on their smallholding, deferring to the judgment of the consultant. Because of the link that the associations provide with credit provision, they serve as a very effective conduit for information about new technologies and have undoubtedly served to speed the adoption of Bt cotton varieties.

We collected survey information on yields, revenue, and pest control costs for the first two years that Bt cotton was widely grown in Mexico (1997 and 1998). The data were collected from the technical consultants working for the association SEREASA, one of the largest of the 14 associations in Comarca Lagunera. In 1997, this association had a total of 638 producers owning 4,789 ha of land. Of this, 2,265 ha were planted to cotton in 1997 and 2,023 ha in 1998—about 12% of the cotton area in the Comarca. The members of the association are probably representative of medium to small landholders in terms of size of holding. The median size holding of SEREASA ejido members was 3.5 ha, while that of SEREASA small landholders was 20 ha. The mean cotton acreage was 15 ha in 1997 and 8 ha in 1998.

The Bt variety NuCOTN 35B was grown on 52% of SEREASA cotton area in 1997; two conventional varieties accounted for 48% of the area. Yields were about the same for both types of cotton, but conventional cotton graded slightly higher, which is reflected in a $65/ton higher average price. As a result, conventional cotton produced nearly $50/ha higher revenue than the Bt variety. Less pesticide, however, was used on the Bt cotton. Conventional cotton averaged 1.57 applications for pink bollworm, but no growers sprayed Bt cotton. Conventional cotton required more than twice as many pesticide applications to control cotton budworm, and slightly more applications for armyworm and other insects. All growers used biological control against cotton bollworms. Bt cotton growers averaged 2.26 fewer total pesticide applications than conventional cotton growers did. Total chemical pesticide costs were $153.91 less for Bt cotton, and total pest control costs, including seed costs, were $92.66 less. The net difference in profitability was a $44.15 advantage for Bt cotton.

Adoption of Bt cotton varieties increased to 72% in 1998, and average Bt yields were 0.29 t/ha higher than for conventional varieties. Lint quality was higher for Bt cotton, giving it a $543.56/ha revenue advantage. An average of two fewer pesticide applications were used on Bt than conventional cotton, and total seed and pesticide costs were $83.19 less. The net profit advantage for Bt cotton in 1998 was $626.74. The large difference in relative profitability of Bt cotton between 1997 and 1998 is likely explained by differences in pest infestation levels. The yield advantage of Bt cotton increases in parallel to infestation levels, and 1997 was a very light year for pink bollworm compared to 1998. By historical standards, even 1998 was not a heavy pink bollworm year.

With more than $600/ha net benefit during years of pest pressure, and slightly higher profits in low pest years, Bt cotton provides growers a valuable insurance against pest infestation. The profit from 1998 would cover technology fees for several years.

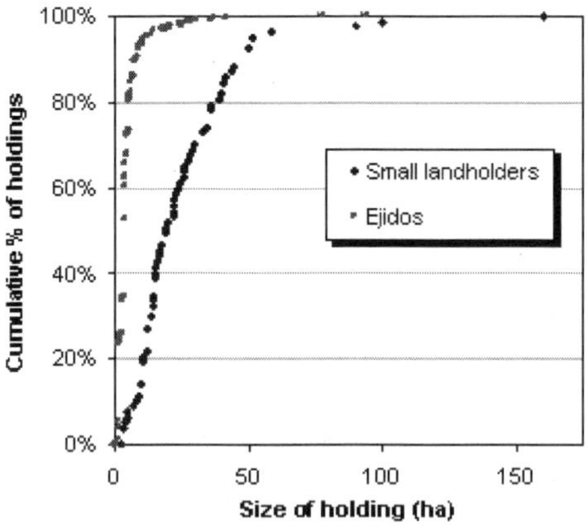

Fig. Size distribution of SEREASA land holdings, 1997.

Table. Estimates of economic surplus distribution in Comarca Lagunera, 1997 and 1998 ($US).

	1997	1998	Average
Conventional seed price per kg	2.21	2.21	2.21
Cost per ha to produce Bt seed	30.94	30.94	30.94
Monsanto/D&PL Bt revenue per ha	101.03	101.03	101.03
Monsanto/D&PL net revenue per ha[a]	70.09	70.09	70.09
Farmer change in variable profit per ha	44.15	626.74	335.40
Bt area in Comarca Lagunera	4,500	8,000	6,250
Monsanto/D&PL total net revenue[a]	315,405	560,720	438,063
Total farmer surplus	198,675	5,013,920	2,096,250
Total surplus[a] produced	514,080	5,574,640	2,534,313
Monsanto/D&PL share of total surplus[a]	61%	10%	17%
Producer share of total surplus	39%	90%	83%

[a] Monsanto/D&PL net revenue calculated before administrative and sales expenses and before any compensation to Mexican seed distribution agents.

BENEFIT DISTRIBUTION BETWEEN MONSANTO/D&PL AND COTTON PRODUCERS

The estimated surplus distribution between Monsanto/D&PL and producers is given inTable 2. After subtracting the estimated cost of seed production, we estimate that Monsanto/D&PL were left with a net revenue of roughly $70/ha. Expenses related to field research, providing technical assistance to farmers, for monitoring contract compliance, or compensation to local seed distribution agents were not subtracted because we do not have this information available. The per-hectare change in variable profit accruing to farmers varied widely between the two years, with an average figure of $335.45. Therefore, for the two years, we estimate that a total of more than $6 million in surplus was produced, of which about 86% accrued to farmers and 14% to Monsanto/D&PL; but again, not all of the amount attributed to Monsanto is true surplus, because some costs were not accounted for.

SEED PRICES, CONTRACTS, AND PROTECTING INTELLECTUAL PROPERTY

When Monsanto/D&PL introduced Bt cotton in 1996 and Roundup Ready soybeans in 1997 in the United States, it also introduced the use of seed licensing contracts, which farmers are required to sign upon seed purchase. The seed contract forbids farmers from saving seed and, in the case of Bt cotton, obligates them to follow a specified resistance management strategy, to refrain from saving seed, to have cotton ginned only at authorized gins, and to contract an entomologist to supervise the farmer's compliance with biosafety standards. Monsanto hires two seasonal field representatives in Comarca Lagunera to spot check cotton fields and to investigate suspected intellectual property (IP) violations. These representatives are equipped with field kits that test for the presence of the Bt gene at a cost of less than $5 per test. The contractually specified penalty for selling seed—120 times the purchase price—appears to be high enough to have prevented large-scale violations, although some transfer among small farmers is rumored in Comarca Lagunera area.

The contracts with gin owners are another legal initiative taken by Monsanto to protect their revenue from Bt cotton. Because cottonseed can only be separated from the lint by ginning, the gins are a logical focal point for Monsanto to capture the Bt cottonseed. Of the 34 cotton gins in Comarca Lagunera that existed in 1990, only 12 remain. In the contract, the gins are offered the opportunity to become "authorized Monsanto cotton gins" by agreeing to refrain from selling or using Bt seed obtained through the ginning process. Given the 96% adoption of Bt cotton in Comarca Lagunera and that the producers' contract calls for ginning only at Monsanto-authorized gins, it is not surprising that all gins have signed Monsanto's contract. The gins also agree to open their facilities and transaction records to inspection by Monsanto. This allows Monsanto to be informed of any producers who have requested their seed back from the gin.

Total revenue from Bt cottonseed sales in Mexico in 2000 was approximately $1.5 million. The price charged for Bt varies by growing region. For example, the technology fee is three and half times higher in Northern Tamaulipas than it is in Southern Sonora, where BBWC problems are the lightest. The differential pricing strategy is based on differences in the marginal value product of Bt cottonseed caused by differences in pest pressure and seed application rates. Monsanto/D&PL have attempted to thwart spatial arbitrage by working with the distributors in each region. Distributors are simply asked to refrain from selling Bt cotton to producers from outside of their region. For example, attempts are made to prevent farmers from buying low-cost seed in Chihuahua for planting in neighboring Comarca Lagunera. This appears to have been effective because of the relatively small acreage involved (about 16,000 ha total in Chihuahua and Comarca Lagunera) and the desire of distributors to maintain good relations with Monsanto/D&PL.

SUMMARY AND CONCLUSIONS

Cotton production in the Comarca Lagunera has undergone a transformation over the past decade. The most notable changes are a reduction in pesticide use and the corresponding reduction in the cost of production. The result has been increased profitability and competitiveness, and a reduction in the risk associated with cotton production failures caused by insect infestations. A number of factors have been important in ushering in this new era in cotton production, including the availability of Bt cotton varieties, reduced cotton acreage, and government support for farm credit and integrated pest management.

Bt cotton varieties are in many ways a nearly ideal innovation for the Comarca Lagunera. The region's victory over the pink bollworm—once the dominant insect pest—would not have been possible without Bt cotton. At an average of less than two total chemical pest control applications per season, cotton has become a low-pesticide crop, benefiting both farmers and residents of the region. Bt cotton varieties have been a tremendously useful tool for the Comarca Lagunera, but because they only protect against a certain spectrum of the pest population, they are not a cure-all for cotton production in all regions, as demonstrated by low adoption in other Mexican states. How relevant for other countries is Mexico's experience with Bt cotton? First, it must be recognized that Mexico is an atypical developing country in several respects. It is large in terms of total agricultural area, the size of its national agricultural research system, and the capacity of its university-based basic research establishment. Mexico also began setting the stage for the use of biotechnology earlier than most countries. It began approving biosafety trials in 1988 and has now accumulated a significant amount of experience with the regulation of transgenics. Cotton production in Comarca Lagunera is also intensive; 95% of cotton is irrigated, yields are high by world standards, infrastructure is well developed, and material, financial, and intellectual inputs are readily available. All of these factors favor the successful adoption of a new technology. Of particular importance in Comarca Lagunera were the key government interventions of credit for financing the purchase of Bt cottonseed combined with technical assistance for small landholders and the implementation of an effective integrated pest management program.

Monsanto has been largely successful in enforcing IP protection in Mexico. The small size of the market and the fact that the Bt gene was introduced into a crop in which seed saving can be monitored through activities at 14 gins and through registers of producer field locations contributed to successful enforcement. Clearly, IP enforcement would be more difficult for other self-pollinating crops, such as wheat, rice, or soybeans, and for crops such as maize, which are grown by more dispersed small farmers. The experience in Mexico suggests, however, that relevant conditions for transferring biotechnology to developing countries through the private sector activities may indeed exist in some situations.

A final point that is worth noting is that despite Mexico's positive experience with Bt cotton, constraints on other biotechnologies do exist. Mexico has about 7.5 million ha of maize, compared to 0.2 million ha of cotton, and would be an attractive market for transgenic maize. However, biosafety testing of transgenic maize has been indefinitely suspended. Therefore, biosafety procedures can be a source of considerable uncertainty, even in experienced countries.

PEST TRADE-OFFS IN TECHNOLOGY: REDUCED DAMAGE BY CATERPILLARS IN BT COTTON BENEFITS APHIDS

Varieties of genetically engineered (GE) cotton, *Gossypium hirsutum* (L.) (Malvales: Malvaceae), that produce Cry proteins derived from *Bacillus thuringiensis* (Berliner) (Bt) are highly resistant against major lepidopteran pests and their widespread planting has led to area-wide population suppression of key pest species . The Cry toxins currently deployed in Bt cotton varieties are not known to cause direct toxic effects to herbivorous pests other than those in the order Lepidoptera . Thus, management of non-lepidopteran pests remains a requirement for the sustainable deployment of Bt-transgenic cotton . Some studies have reported increased populations of sap-feeding herbivores, such as aphids and leafhoppers, as well as plant-feeding hemipterans such as plant bugs (Miridae) and stink bugs (Pentatomidae), in Bt cotton . In the case of plant-feeding hemipterans, increased damage to Bt cotton has been linked to the reduced use of broad-spectrum insecticides to control pest Lepidoptera . However, increased abundance of non-lepidopteran herbivores in Bt cotton was also reported from some studies that excluded chemical insecticide treatments on both Bt and non-Bt cotton , suggesting that reduced insecticide use alone is an insufficient explanation for some emergent pests in cotton.

Cotton produces a range of closely related terpenoids (e.g. gossypol) that act as anti-feedants and possess insecticidal properties . The levels of these defensive compounds are systemically increased by the plant damage caused by tissue feeders, such as lepidopteran larvae . The intensity of the damage affects the levels of induced terpenoids . This can have cascading effects on the performance of tissue feeders and sap-feeding herbivores, such as aphids , which by themselves do not appear to induce this plant resistance .

We demonstrate here in a glasshouse and field study that the reduced damage caused by larvae of *Heliothis virescens* (Fabricius) (Lepidoptera: Noctuidae) on Bt cotton leads to a reduced induction of cotton terpenoids, which enhances the crop's susceptibility to a non-target pest, namely the cotton aphid *Aphis gossypii* (Glover) (Hemiptera: Aphididae). This indirect, plant-mediated effect of Bt cotton on non-target species has not been previously considered and may constitute an additional mechanism that benefits non-target herbivores.

MATERIAL AND METHODS

GLASSHOUSE EXPERIMENTS

(i) Plants: Two varieties of GE cotton plants, which were provided by Monsanto Company (St. Louis, USA), were used for the experiments. One variety ('Bt cotton'), Deltapine DPL143B2 RF (event: MON15985 × MON88913), expresses two Bt toxins (Cry1Ac and Cry2Ab, Bollgard II) and carries an herbicide-tolerance trait (targeting glyphosate). The second variety ('non-Bt cotton'), Deltapine DPL147 RF (event: MON88913), has a similar genetic background but only contains the herbicide-tolerance trait. For insect rearing, the closely related non-GE variety Deltapine DPL491 was used. Plants were grown in 3 l plastic pots containing heat-sterilized humus-rich soil. At

planting, 15 mg of the slow-release fertilizer Osmocote (16% N, 11% P_2O_5, 11% K_2O; Scotts UK Professional, Bramford, UK) was added to each plant. Subsequently, the plants were fertilized weekly using 10N : 10P : 8K at 20 ml l”^{1} and were watered daily with water containing no fertilizer. Plants were enclosed in gauze cages (height, 71 cm; diameter, 35 cm; mesh-width, 0.264 mm) to protect them from glasshouse pests. Bt and non-Bt cotton plants were used when they had four fully expanded true leaves.

(ii) Insects: A colony of *A. gossypii* was obtained from Syngenta (Stein, Switzerland) and reared on four- to eight-week-old cotton plants (Deltapine DPL491). The *H. virescens* larvae used in the experiments were regularly obtained from Syngenta.

(iii) Experiments: The glasshouse experiments were conducted at 25°C ± 5°C, relative humidity 70 ± 10% and 16 L : 8 D long-day conditions.

Plants were either infested with a single *H. virescens* larva (3rd instar) that was caged on the youngest fully developed leaf for 7 days in a gauze bag, or left uninfested (control). *Heliothis virescens* larval weight was recorded before placing the larva on the plant and after 7 days to calculate the weight gain. In total, 27 plants were initially tested per treatment. Plants damaged accidentally during the experiment or found to be infected by any glasshouse pest were removed, resulting in n = 15–21 per treatment.

After the 7 days, 20 *A. gossypii* (mixed stages) were transferred to the youngest fully developed leaf of each plant (not identical to the leaf on which the larva was released). The aphids were enclosed in a clip-cage that was removed after 24 h. After 14 days, the aphid numbers were recorded, and all aphids were transferred into vials with 70 per cent ethanol. Subsequently, hind-tibia lengths were measured for 10 randomly chosen adults (fewer if 10 were not available) from each plant.

After termination of the experiment, leaf damage caused by *H. virescens* larvae was recorded. For this, the damaged leaf from each plant was collected and photographed and the damaged leaf area was measured with IMAGEJ v. 1.42 software.

A separate set of Bt and non-Bt cotton was treated as described earlier, with one group from each plant type being infested with *H. virescens* while the second remained uninfested (10 plants per treatment). After 7 days, the youngest fully developed leaf and the oldest true leaf from each plant were collected, shock-frozen in liquid nitrogen, stored at “80°C, and later freeze-dried for high-performance liquid chromatography (HPLC) analyses.

To test whether the Bt and non-Bt cotton plants respond similarly to a known resistance inducer, plants were treated with jasmonic acid (JA) . A 100 μM JA solution was prepared by dissolving JA in 100 per cent ethanol. The JA–ethanol solution was then diluted in water 4 : 1000 (JA–ethanol : water). Cotton plants (fourth-full-leaf stage) were watered with 25 ml of the JA solution. The water was applied by pouring it on the stem of the plant, thereby allowing uptake of JA by above- and below-ground tissue. The control plants were treated with an ethanol : water (4 : 1000) solution. Each treatment included 10 plants. After 7 days, the youngest fully developed leaf and the oldest true leaf from each plant were collected, shock-frozen in liquid nitrogen, stored at “80°C and later freeze-dried for HPLC analyses.

FIELD EXPERIMENT

The field experiment was conducted at the USDA-ARS Coastal Plains Experiment Station, Tifton, GA (USA). Bt cotton (Deltapine DPL143B2 RF) and non-Bt cotton (Deltapine DPL147 RF) were planted on 30 April 2009, with three plots (27 × 49 m) of each variety distributed in a randomized complete block design over an area of *ca* 3.5 ha. The rest of the field was planted with plots of soybean (Stoneville 78-G7, Maturity Group 7), peanut (Georgia green) and cotton. Plots were separated by 6 m of bare, tilled soil. Aldicarb (Temik) was applied to furrows at 5.6 kg ha”1 at planting to protect the plants from early season pests. Otherwise, the crop was grown and managed according to agronomic practices recommended by the Georgia Cooperative Extension Service except that no insecticides were applied after planting. Glyphosate (Roundup Ultra Max) was applied on 29 May, and the plant growth regulator Mepiquat chloride (Compact) was applied on 25 June. Plants were fertilized with liquid nitrogen in mid-June. The plots were irrigated as required. Data from two plots (one Bt and one non-Bt) were not considered for the analyses because they contained substantial numbers of fire ants, which probably affected herbivore abundance.

When plants were four to five weeks old, 30 randomly selected plants in each of the four plots were tagged with coloured flags and ribbons to be able to resample the same plants. Ten of these 30 plants in each plot were artificially infested with *Spodoptera exigua* (Hübner) (Lepidoptera: Noctuidae), a species that infests cotton fields early in the season compared with other Bt-sensitive herbivores and is well known to induce the production of cotton terpenoids . Each week, the same 10 plants in each plot were infested with three 2nd instar *S. exigua* larvae. The larvae were caged onto a randomly selected leaf from the middle third of the plant. Each week plants were infested on a new leaf.

Between 1 June and 20 July, the aphid population was counted twice each week on the youngest fully developed leaf and on a random leaf from the middle third of each of the 120 tagged plants, i.e. 40 plants that were artificially infested with *S. exigua* and 80 plants that were exposed only to natural infestation by herbivores. Once per week, starting on 10 June, the damage to the foliage of every plant was recorded in a non-destructive way by photographing every damaged leaf and measuring the area of removed tissue using IMAGEJ v. 1.42 software.

For chemical analyses, the youngest fully developed leaf of all tagged plants was collected on 20 July 2009. Samples were placed immediately on ice in the field and were subsequently transferred to "80°C and later freeze-dried for terpenoid quantification.

TERPENOID QUANTIFICATION

Samples were prepared and extracted according to the procedure of Benson *et al.* . An 8–10 mg sample of freeze-dried and ground leaf tissue per plant was extracted with 1 ml of a mixture of acetonitirile (Multisolvent HPLC grade, Scharlau, Sentmenat, Spain), water (purified by a Gradient A10, Millipore, Billerica, USA) and 85 per cent phosphoric acid (Fluka, Buchs, Switzerland) (80 : 20 : 0.1) for 3 min in an ultra-sonicator at room temperature.

After 3 min of centrifugation (8000*g*), the extract was directly transferred into glass vials for separation and detection on an liquid chromatography system (1090 Series II, Hewlett-Packard, Palo Alto, USA). All substances to be analysed in the plant extracts were baseline separated at 40°C on a Varian Polaris Amide C-18 column (150 × 2.0 mm, 3 μm) equipped with a precolumn (C18, 4 × 3.0 mm, Supelco Security Guard System) and were detected with a single wavelength absorbance detector at 272 nm.

The following elution gradient was applied: 0 min: 60 per cent B (40% A); 2 min: 60 per cent B; 20 min: 70 per cent B; 21 min: 100 per cent B; 25 min: 100 per cent B; 26 min: 60 per cent B; 35 min: 60 per cent B. Eluent A consisted of MilliQ-water, and eluent B consisted of acetonitrile. Eluent A was acidified with 0.1 per cent trifluoroacetic acid (greater than or equal to 99%, Riedel de Haën, Seelze Germany) to pH = 2.5 (based on). The flow rate of the mobile phase was 0.125 ml min^{-1}, and a 10 μl sample was injected.

Gossypol (e" 95%, Sigma, St. Louis) was identified by comparing the retention time in the extract with the retention time of standard solutions. The order of elution of hemigossypolone and the heliocides 1–4 was ascertained from previously published chromatograms and analytically confirmed by mass spectrometry. Terpenoid concentration was expressed in terms of gossypol equivalents .

STATISTICAL ANALYSES

(i) Glasshouse experiments: Plant damage caused by *H. virescens* was analysed using the Welch *t*-test for non-homogeneous variances. Aphid abundance was compared among treatments using one-way analysis of variance (ANOVA).

Means were subsequently separated using the Tukey Honestly Significant Difference (HSD) test. Data were log-transformed to meet the ANOVA assumptions when necessary. Aphid tibia length was analysed using Kruskal–Wallis ANOVA.

(ii) Field experiment: Data obtained from plants of the same plot were pooled to adjust for dependency. Separate analyses were conducted for naturally infested plants and plants that were artificially infested with *S. exigua*. Aphid data from the field and damage levels of plants in the plots were analysed by repeated measures ANOVA. Differences of aphid abundance on specific sampling dates were separated using Fisher's Least Significant Difference (LSD) test. Correlations between aphid counts from a specific date and plant damage, as well as total terpenoid concentration of plants, were evaluated using Kendall's τ rank correlation. For the correlation the data were not pooled by plot.

(iii) Terpenoid analyses: For analyses of cotton terpenoids in glasshouse grown plants, one-way ANOVA was used. Means were subsequently separated using the Tukey HSD test. Separate analyses were conducted for young and old cotton leaves. Data were log-transformed, if required. Terpenoid concentrations in leaf samples from the field were compared between Bt and non-Bt plots using Student's *t*-test.

RESULTS

GLASSHOUSE EXPERIMENTS

When exposed to a single *H. virescens* larva for 7 days, Bt cotton plants remained nearly undamaged (average leaf area consumed ± s.e.: 0.12 ± 0.02 cm^2; $n = 12$) compared with non-Bt cotton plants (31.79 ± 5.66 cm^2; $n = 13$; *t*-test: $t =$"$5.37, p < 0.0001$). No larvae ($n = 18$) were found alive after 7 days on Bt cotton, but all larvae on the non-Bt cotton survived and gained considerable weight (average weight gain ± s.e. per larva: 136.4 ± 34.4 mg; $n = 15$).

The constitutive expression of foliar terpenoids did not differ between undamaged Bt and non-Bt cotton plants in their oldest and youngest fully developed leaves ($p > 0.05$). This was true for the four terpenoid classes that we analysed: gossypol, hemigossypolone, heliocides 1 + 4 and heliocides 2 + 3. Herbivory by *H. virescens* induced terpenoid synthesis by the non-Bt plants: terpenoid levels were significantly increased in young leaves of herbivore-challenged non-Bt plants compared with Bt cotton and undamaged plants ($F_{3,36} = 27.98, p < 0.0001$). This increase was significant for three of the four major classes of cotton terpenoids that were analysed. Terpenoid levels in the oldest true leaves were unaffected by *H. virescens* feeding and did not differ among treatments.

The application of JA confirmed that terpenoid production is equally inducible in Bt and non-Bt cotton plants ($F_{3,36} = 6.22, p < 0.001$). This increase was significant for the four major classes of cotton terpenoids that were analysed. Terpenoid levels in the oldest true leaf were not inducible and did not differ among treatments. To investigate how the Bt-mediated differences in induction of terpenoids by *H. virescens* affect Bt-insensitive herbivores, we measured the population growth of *A. gossypii* on Bt and non-Bt cotton in the glasshouse. Uninfested Bt and non-Bt cotton plants were equally suitable as aphid hosts (p = 0.99). When cotton plants had previously been exposed to *H. virescens*larvae, however, aphids were significantly more abundant on the Bt cotton than on the non-Bt cotton plants ($F_{3,69}$ = 6.22; p < 0.001). The median of hind-tibia length (which was used as an indicator of aphid size) ranged from 0.28 to 0.30 mm and did not differ among the different plant treatments (□$^2_6$ = 4.11, p = 0.66).

FIELD EXPERIMENT

Bt plants experienced significantly less damage than non-Bt plants, when plants were artificially infested with *S. exigua* (repeated measures ANOVA; plant type: $F_{1,2}$ = 997.56, p = 0.001; plant type × time interaction: $F_{5,10}$ = 131.10; p < 0.0001). As a consequence, the total terpenoid concentrations measured in leaf samples collected in late July were lower in infested Bt plants (12 149 ± 260.1 ng g^{-1} dry weight (dw) leaf) than in infested non-Bt plants (14 949 ± 542.9 ng g^{-1} dw leaf) (*t*-test; t_2 = 4.65, p = 0.043). This difference was mainly because of an increased gossypol concentration.

In the case of naturally infested plants, however, the damage between Bt and non-Bt plots did not differ significantly (repeated measures ANOVA; plant type: $F_{1,2}$ = 1.01, p = 0.42; plant type × time interaction: $F_{5,10}$ = 1.76; p = 0.21). As a consequence, the total terpenoid concentrations measured in leaf samples collected in late July did not differ between plant types (Bt cotton: 8990 ± 1405.6 ng g^{-1} dw leaf; non-Bt cotton: 9815 ± 19.2 ng g^{-1} dw leaf) (*t*-test; t_2 = 0.59, p = 0.62).

Aphids began to appear in mid-June. Aphid density peaked around mid-July, after which the population dropped. A quick collapse of the *A. gossypii* population on cotton is a typical pattern in the Southeastern USA and is caused by epizootics of the entomopathogenic fungus *Neozygites fresenii* (Nowakowski) (Entomophthorales: Neozygitaceae) . During the observation period, aphid populations developed differently in plots with Bt cotton than in those with non-Bt cotton when the plants were artificially infested with *S. exigua* (repeated measures ANOVA; plant type: $F_{1,2}$ = 3.14; p = 0.22; plant type × time interaction: $F_{12,24}$ = 4.52; p = 0.001) or when naturally infested (plant type: $F_{1,2}$ = 9.12; p = 0.09; plant type × time interaction: $F_{12,24}$ = 2.7; p = 0.024). Aphids were significantly more abundant on naturally infested Bt plants compared with non-Bt plants on 13, 16 and 20 July. Likewise, aphids were more abundant on artificially infested Bt plants on 13 and 20 July. Terpenoid concentrations in leaf samples collected on 20 July were not positively correlated with leaf damage recorded on 15 July (naturally infested plants: □ = 0.033; p = 0.67, n = 79; *S. exigua*infested plants: □= 0.082; p = 0.47, n = 39). The cumulative aphid abundance until 20 July was negatively correlated with the plant damage on 15 July for naturally infested (□=”0.178; $P = 0.020, n = 79$) and *S. exigua* infested plants (□=”0.22; $p = 0.051, n = 39$). No correlation between aphid populations and terpenoid concentrations of the plants was detected (naturally infested plants: □= “0.10; $p = 0.20, n = 80$; *S. exigua* infested plants: □=”0.09; $p = 0.39, n = 40$).

DISCUSSION

Our results show that effective suppression of target herbivores by Bt cotton translates into a decrease in the level of induced terpenoids, leaving plants more susceptible to the cotton aphid*A. gossypii*, a herbivore not targeted by the Bt toxins. This effect was very pronounced under protected glasshouse conditions but also visible in the field, although to a much smaller degree.

That prior herbivory renders plants more resistant to subsequent attack has been well established for a number of non-GE systems and is commonly referred to as indirect plant-mediated competition. Some studies could link plant-mediated competition to the well-documented induction of plant secondary metabolites in response to herbivory . Wild radish plants damaged by *Pieris rapae* (L.) (Lepidoptera: Pieridae) show increased glucosinolate levels corresponding with adverse effects on Lepidoptera, aphids and a leafminer . In cotton, a number of studies have linked secondary metabolite induction to reduced fitness for the inducing herbivore . Induction of terpenoids in cotton has also been reported in relation to plant-mediated competition, both between root- and shoot-feeding herbivores , and between the spider mite *Tetranychus urticae* (Koch) (Acari: Tetranychidae) and the fungal pathogen*Verticillium dahliae* (Kleb.) (Ascomycetes: Hypocreales) sharing cotton as a host plant . In addition to conferring resistance against insect herbivores, cotton terpenoids act on a broad range of other organisms, including pathogens and nematodes . The reported differences in induced terpenoid levels between Bt and non-Bt cotton are, therefore, likely to affect these organisms as well.

Emergent pest problems in crops expressing Bt toxins have been attributed to the reduced use of broad-spectrum insecticides , reduced competition from pests targeted by the Bt trait, or unintended transformation-related effects. In the latter case, the transformation process may generate inadvertent changes in the plant, rendering it more susceptible to non-target pests, as has been suggested for aphids on Bt-transgenic maize . However, we found no evidence of such unintended transformation-related effects in our glasshouse experiment because the increases in aphid numbers and the size of the aphids in the absence of caterpillars were similar on the Bt-transgenic cotton and its non-transformed comparator, confirming earlier studies with other Bt cotton events . Furthermore, the Bt and non-Bt cotton plants did not differ in their constitutive levels of terpenoid expression, and both plant types showed equal levels and comparable patterns of terpenoid induction in response to treatment with JA.

Indirect plant-mediated herbivore–herbivore competition can be based on a range of herbivore-induced changes other than induced secondary metabolites. These include changes in plant morphology or overall resource quantity . There is evidence that release from such exploitative (or resource) competition in Bt cotton may also contribute to emergent pest development. Protection from lepidopteran damage significantly reduces damaged squares and flowers in Bt cotton , which benefits stink bugs and plant bugs that feed on these structures . In addition, there is evidence that the herbivores also profit from reduced physical interaction with the Bt-targeted caterpillars .

In the field, the results from the glasshouse experiment could only partly be confirmed. Artificial infestation with *S. exigua* larvae caused larger damage to non-Bt plants compared with Bt plants and consequently led to a higher terpenoid concentration in the non-Bt plants. This difference in damage and terpenoid concentration, however, was not visible for naturally infested Bt and non-Bt cotton. Nevertheless, the development of the aphid population differed between Bt and non-Bt plants in both cases, for artificially and naturally infested plants. The fact that cumulative aphid populations were not correlated with the amount of terpenoids measured in the cotton plants at the termination of the experiment indicates that, under field conditions, factors other than terpenoids also impact the aphid populations. These factors include other chemical compounds or morphological changes of the plant that affect the attractiveness or nutritional quality of the plant , and activity of natural enemy species, especially coccinellid species later in the season . Additionally, a lack of correlation could arise from terpenoid induction changes over time , making it difficult to identify the point in time at which the altered terpenoid levels translate into detectable changes in aphid populations.

There was also a lack of correlation between leaf damage and terpenoid concentration in the field, which might be caused by the large complex of pest species that attack cotton. Cotton plants can react very differently to herbivory by even closely related pest species and depending on the part of the plant on which they feed . Therefore, the measurement of leaf damage alone might not be sufficient to adequately predict the degree of induced resistance in a cotton plant.

Herbivore–herbivore interactions mediated by induced plant compounds have not been previously considered as a mechanism to explain the increased susceptibility of Bt-transgenic crops to pests that are insensitive to the Bt trait. Interactions between introduced insecticidal Cry toxins and inherent resistance mechanisms have only been discussed in the context of improving the insect resistance of Bt cotton and reducing the risk of resistance evolution in the target pests by increasing the fitness costs associated with resistance to Bt . The interaction between introduced insecticidal Cry toxins and induced levels of secondary metabolites demonstrated in our study may also help explain results from studies showing inconsistencies between levels of Cry toxin in the plant tissues and survival of Bt-susceptible herbivores .

Even though the aphids in our study performed better on Bt-transgenic than on caterpillar-damaged non-Bt cotton plants, aphid outbreaks are not more common in commercial Bt cotton fields than in non-Bt cotton fields. This may be because herbivore populations may be suppressed by enhanced survival of natural enemies in Bt cotton as a result of reduced use of insecticides against target Lepidoptera or because generalist predators feed more on aphids in the absence of caterpillars. This would generate increased top-down control of Bt-insensitive pests which could offset enhanced aphid performance . For mirids, stinkbugs and other herbivores that lack effective natural enemies, this counteracting mechanism is apparently insufficient to prevent non-target pests from developing pest status .

In addition to being relevant to pest control with Bt crop plants, the mechanism reported here is likely to be relevant to other control methods that provide a comprehensive suppression of one group of herbivores. The common phenomenon that (secondary) pest levels increase following insecticide treatments could also be partly attributed to lower levels of induced resistance resulting from the reduced herbivory by primary pests. The plant resistance-mediated interaction reported here probably pertains to a broad range of agricultural crops expressing induced resistance. It follows that these findings could have wide-reaching implications for interactions between plants and associated organisms.

FURTHER READING

- "Cotton: The Biography of a Revolutionary Fiber" by Stephen Yafa, Publisher: Penguin USA, ISBN 978-0143037224, ISBN 0143037226
- Brown, D. Clayton. *King Cotton: A Cultural, Political, and Economic History since 1945* (University Press of Mississippi, 2011) 440 pp. ISBN 978-1-60473-798-1
- Ensminger, Audrey H. and Konlande, James E. *Foods and Nutrition Encyclopedia,* (2nd ed. CRC Press, 1993). ISBN 0-8493-8980-1
- USDA – Cotton Trade
- Moseley, W.G. and L.C. Gray (eds). *Hanging by a Thread: Cotton, Globalization and Poverty in Africa* (Ohio University Press and Nordic Africa Press, 2008). ISBN 978-0-89680-260-5
- Smith, C. Wayne and Joe Tom Cothren. *Cotton: origin, history, technology, and production* (1999) 850 pages
- True, Alfred Charles. *The cotton plant: its history, botany, chemistry, culture, enemies, and uses* (U.S. Office of Experiment Stations, 1896) online edition
- Yafa, Stephen H. *Big Cotton: How A Humble Fiber Created Fortunes,*

Wrecked Civilizations, and Put America on the Map (2004) excerpt and text search

REFERENCES:

1. *Cotton for Nonwovens"*: A Technical Guide, Cotton Incorporated.
2. Lawrence H. Shaw; " *Cotton's Importance in the Textile Industry*", Symposium, Lima, Peru, May 12, 1998
3. Tortora, P.G., Collier, B.J.*: "Understanding Textiles"*, 5th edition, Prentice-Hall, 1997
4. Kadolph, S.J., Langfold, A.J.: *Textiles*, 8th edition, Prentice-Hall, 1998
5. www.freedoniagroup.com
6. U.S. cotton Market, Monthly Economic Letter, Cotton Incorporated, Market Research, Sep 15th, 1997
7. *"The Classification of Cotton"*, USDA Agricultural Marketing Service, cotton Division, Agricultural Handbook 566, September 1995,br>
8. Shaw, C., Eckersley, and F.: " cotton", Sir Isaac Pitman & Sons Ltd., London, 1967
9. Duckett, K.E.: "*Surface Properties of Cotton Fibers*", Surface Characteristics of Fibers and Textiles, edited by M.J.Schick. "*Fiber Science Series*", Marcel Dekker, Inc. 1975, p 67,br>
10. Matthew's "*Textile Fibers, Their Physical, Microscopic and Chemical Properties*", edited by Herbert R. Mauersberger, 6th edition, John Wiley & Sons, Inc., 1954
11. Webster's Third New International Dictionary, edited by Phillip Babcock Dove, G. & C. Merriam Company, 1963.
12. Gordon Cook, J.: " *Handbook of Textile Fibers, Part I. Natural Fibers*", Merrow Publishing Co. Ltd., 1968
13. Lawrence H, Shaw: "*Cotton's future trend* ", 28th Annual Congress of the Commercial Cotton Growers of Zimbabwe, June 5, 1996
14. M. Rafiq chaudhry: "*Harvesting and ginning of cotton in the world*", Technical information section, International Cotton Advisory Committee, Washington, D. C. 1997
15. National Agricultural Statistics Service (NASS), Agricultural Statistics Board, U.S.Department of Agriculture. http://usda.mannlib.cornell.edu/reports/nassr/field/pcp-bba/acrg0699.txt Released June 30, 1999
16. 16.M Rafig Chaudhry: "*Cost of Producing a Kilogram of cotton*", Technical information section, International Cotton Advisory Committee, Washington, D. C. 1997
17. US Patent5634243, Ripley; W. G. June 3, 1997
18. http://www.mini-magic.com/mini/fabric.htm
19. Kermit E. duckett: "*Color grading of cotton-measurement*", Beltwide cotton conference, Orlando, Jan. 5-8, 1999
20. J. Brandrup; E. H. Immergut; "*Polymer Handbook*", 1989

21. M. Dean Ethridge, 57th Plenary Meeting of International Cotton Advisory Committee, Santa Cruz, Bolivia, Oct. 12-16, 1998
22. H. Charles Allen, Jr.; "*Cotton in Absorbent Cores*", Nonwovens World, August-septembet, 1999, 71-78
23. Mark D. Lange, "*Cotton Markets in the Crystal Ball*', Textile Month, June, 1998, 37-40
24. ATI Special report, "Outlook for U.S cotton 1999", ATI, May 1999, 140-156
25. Cotton Fibers: http://www.nonwovens.com/facts/technology/fibers/cotton.html
26. Judith M. Bradow, etc; "*Quality Measurements*", The Journal of Cotton Science, 1:48-60, (1997)
27. P. Bel-beiger, etc; "*Textile Technology, Cotton/Kenaf Fabrics*: A Viable Natural Fabrics", the Journal of Cotton Science, 3:60-70, (1999)
28. "*A Guide to Fibers For Nonwovens*", Nonwoven Industry, June 1999, 60-82
29. "*Readers Service, Natural Cotton Fiber*", Nonwoven Industry, Jan. 1999, 74
30. B. Xu and C. fang; "*Clustering An3alysis For Cotton Trash Classification*", Textile Research Journal, 69(9), 656-662, 1999
31. Y. Cai, etc; "*A New Method for Improving the Dyeability of cotton with reactive Dyes*", Textile Research Journal, 69(6), 440-446, 1999

6

Cotton and Fibre Industry

INTRODUCTION TO COTTON

Several species of the genus Gossypium provide what we call cotton. The fine fibrous hairs that occur on the seeds constitute the raw material. These hairs are flattened, twisted and tubular. They compose the lint, floss or staple. Their length and other qualities vary with the different varieties. The plant is a perennial shrub or small tree naturally, but under cultivation it is treated as an annual. It branches freely and grows to a height of 4-8 ft. Cotton thrives in sandy soil in humid regions that are near water. This environment is typified in the southern United States and in the river valleys of India and Egypt. Cotton matures in 5-6 months and is ready to harvest soon after.

KINDS OF COTTON

Hundreds of varieties have been developed from wild ancestors or produced by breeding during the long period of cultivation. Varieties differ in fibre character as well as other morphological features. Cotton is a difficult group to classify and the exact number of species is subject to argument. Cultivated cottons of commercial importance are usually referred to one or another of four species: Gossypium barbadense and G. hirsutum in the Western Hemisphere and G. arboreum and G. herbaceum in the Eastern Hemisphere.

Gossypium Barbadense

Probably originated in tropical South America. The flowers are bright yellow with purple spots. The fruit, or boll, has three valves, and the seeds are fuzzy only at the ends. Two distinct types exist:

Sea-Island Cotton

This type has never been found in the wild as it was already being cultivated at the time of Columbus. It has fine, strong and light cream-colored fibres that are regular in the number and uniformity of the twists and they have a silky appearance. These characteristics are valuable and sea-island cotton was formerly in great demand for the finest textiles, yarns, and lace and spool cotton. Sea-island cotton was brought to the United States from the West Indies in 1785. The finest types were developed on the islands off the South Carolina coast and adjacent mainland. Here strong and firm stables of two inches or more in length were produced. Another form of sea-island cotton was grown along the coast in Georgia to Florida and in the West Indies and South America. This has a staple of 1.5-1.75 in. in length. The yield of sea-island cotton was lower than other kinds of cotton, but this was compensated for by the greater value of the fibre. The boll weevil almost completely eradicated production of sea-island cotton before control measures were discovered.

Egyptian Cotton

This cotton is grown in the Nile basin of Egypt where it was introduced from Central America. The plant is similar in appearance to sea-island cotton and is believed to be a hybrid. However, the staple is brown in colour and shorter. Its length, strength, and firmness make this cotton suitable for thread, undergarments, hosiery, and fine dress goods. Egyptian cotton was brought to the United States in 1902 as an experimental crop and 10 years later it was recommended to farmers in the semiarid regions that were under irrigation. It was then grown in the western states of California, New Mexico and Arizona. Repeated selection and breeding resulted in the development of new strains of whichPima Cotton is of highest quality. Gossypium hirsutum is a native American species that was grown by Pre-Columbian civilizations. It is usually called Upland Cotton, and is the easiest and most economical kind of cotton to grow. It constitutes the greater part of the cultivated cotton of the world. The flowers are white or light yellow and unspotted. The bolls are four- or five-valved, and the seeds are covered with fuzz. Upland cotton thrives under a variety of conditions but does best in a sandy soil with abundant moisture during the growing and fruiting season and dryness during the time of boll opening and harvest as well as a temperature of 60-90 deg.

Fahrenheit. The northern limit of economic growth is 37 deg. N. Lat. The Cotton Belt of the southern United States grows mostly upland cotton. The fibres are white with a wide range in staple length (5/8ths to 1.3/8ths in.). There are over 1210 named varieties, many of which were developed through breeding experiments. The species probably originated in Guatemala or southern Mexico and spread northward to its present limits in North America. A well-marked variety, often recognized as a distinct species, occurs in the West Indies and along the dry coastal areas of South America as far as Ecuador and Brazil. Another variety occurs in Central America, northward along the Gulf of Mexico to Florida and the Bahamas and in the Greater Antilles. Gossypium arboreum is the perennial tree cotton of Africa, India and Arabia. It was most likely the first to be used commercially, but production is now confined to India. The staples are coarse and very short (3/8ths to 34 in. long), but they are strong. Gossypium herbaceum is the principal cotton of Asia. It was grown in Indian in ancient times and continues to be used locally there and in Iran, China and Japan. Its chief use is for fabrics, carpets and blankets and is often blended with wool. There are additionally several wild species of Gossypium in some tropical and subtropical areas.

COTTON INDUSTRY

Cotton used to be an expensive material because it was difficult to remove the fibres from the seed. The cotton gin developed by Eli Whitney in 1793 changed this situation and a revolution of the industry was started. Cotton then assumed a very prominent position in world commerce. The economics of cotton has had a profound effect on both the producing and purchasing nations. It is well accepted that slavery was perpetuated in America because of this crop. There are several steps necessary in the preparation of raw cotton fibre in order to prepare it for the textile industry. These operations involve ginning in either a saw-tooth or a roller gin, baling, transporting to the mills, picking to remove any foreign matter and delivers the cotton in a uniform layer, lapping where three layers are combined into one, carding, combing and drawing where the short fibres are extracted and the others straightened and evenly distributed, and finally twisting the fibres into thread.

COTTON USES

Cotton is used either by itself or in combination with other fibres in the manufacture of all types of textiles. Unspun cotton is extensively used for stuffing purposes. Treating the fibres with caustic soda, which imparts a high luster and silky appearance, makes Mercerized Cotton. Absorbent Cotton consists of fibres that have been cleaned and from which the oily covering layer has been removed. It is almost pure cellulose and makes up one of the basic raw materials of various cellulose industries. A noteworthy advance in the cotton industry was the utilization of what were formerly waste products. In the early stages of the industry the cotton seed along with its fuzzy covering of short hairs or linters was discarded. However, all parts of the plant are now conserved to yield products that are valuable. The stalks contain a fibre that can be used to make paper or fuel and the roots possess a crude drug. The seeds are used for oil extraction and for livestock feed. The linters give wadding, stuffing for pads, cushions, pillows, mattress, etc; absorbent cotton; low grade yarn for twine, ropes and carpets; and cellulose. The hulls are also livestock feed; fertilizer; lining oil wells to prevent cave-ins of the sides; as a source of Xylose, a sugar that can be converted into alcohol or various explosives and industrial solvents. The kernels yield an important fatty oil, cottonseed oil; and oil cake and meal are used for fertilizer, livestock feed, and flour and as a dye.

TYPES OF PLANT FIBERSTYPES OF PLANT FIBERS

Cotton: Cotton plants are related to hollyhock and hibiscus and are grown in warm climates. Cotton fibres are up to 95 percent pure cellulose and derive from. The white, fluffy fibres grow around the plant's seed pod, or boll. Once mature, the bolls are picked and a cotton gin is used to separate the fibres from the seed. Depending on the variety, cotton fibres vary in staple length and softness, but all are relatively durable. Cotton doesn't insulate well, but is extremely absorbent; fibres can absorb more than twenty times their weight in water, and the water will evaporate quickly. Because of the inelasticity of the fibre, items made from cotton will relax or grow over time, and will need to be reshaped after washing. Hemp: Hemp is the coarse, durable bast (skeletal or structural) fibre of the high-yield Cannibis sativa plant (the fibre does not contain the chemicals found in the leaves cultivated for medicinal or recreational drug use). Used historically for items such as ropes and ship sails, it also makes extremely strong paper that doesn't require harsh chemical processing to produce. Hemp fabrics can be machine washed and dried, and they soften considerably after laundering.

Jute: One of the strongest natural fibres, jute is the bast fibre of the Corchorus olitorius and Corchorus capsularis plants grown in monsoon climates. Cultivated in India since ancient times, jute is resistant to heat and fire damage and is biodegradable. Due to its coarseness, jute fabric is used primarily for items such as sacs, rugs, upholstery, twine and rope. Sisal: Originating in the Yucatan Peninsula of Mexico, sisal, like jute, is most often used in ropes, twines, rugs, and other relatively coarse items requiring strength. Flax: Grown in a variety of climates and soil conditions since ancient times, flax plants yield linen fiber. Linen is a strong, coarse fibre that drapes well and, like other cellulose fibres, is durable but lacks elasticity. The more it is worked with and used, the softer and more lustrous it gets. Ramie: Ramie is a bast fibre that is mechanically scraped from the bark of the plant, which is a member of the nettle family. Cultivated primarily in China, the fibre has been used in textiles for over 5,000 years, and resists bacteria and fungal damage. Three times more porous than cotton, it breathes extremely well and is thus suited to hot-weather garments. It takes dye well and is often blended with other fibres for yarn construction.

SEED AND FRUIT FIBRES SEED AND FRUIT FIBRES

Only three seed and fruit fibres have commercial importance:

1. Cotton
2. Kapok
3. Coir.

Cotton is the most widely used of all fibre plants. Cotton fibres are unicellular hairs (trichomes) that emerge from the seed coat after fertilization.

Cotton fibres exhibit two forms: the long "lint" fibres that are twisted into thread and woven into fabrics, and short "fuzz" fibres that are used for batting, felts, and paper production. Single-celled lint fibres grow rapidly and expand approximately 2,500-fold (from 0.020 to 50 millimeters) during maturation. Cultivated for over four thousand years, cotton fibre has historically been obtained from two diploid species (G. arboreum, G. herbaceum) native to Africa and Asia, and two tetraploid species (G. barbadense, G. hirsutum) native to the Americas.

Presently, the tetraploid species account for nearly all of the worldwide production of cotton fibre. Unlike the seed hair of cotton, kapok fibres are produced by the inner surface of fruit pods (capsules) from the silk cotton tree Ceiba pentandra. Kapok fibres reach 20 millimeters in length at maturity, and are waxier and one-sixth the weight of cotton fibres. These properties make kapok difficult to spin; however, its water resistance, light weight, and resilience make kapok an excellent waterproof material for upholstery and life-preservers.

Coir fibre is made from the husk (mesocarp) of fruits from the coconut palm Cocos nucifera. To produce coir, the husks are retted for up to a year, then beaten to separate individual fibres. The cleaned fibres can be spun into coarse yarns for use in ropes and matting, or for bristles in brushes and brooms. Plants yielding fibres have been second only to food plants in their usefulness to humans and their influence on the furthering of civilization. Primitive humans in their attempts to obtain the three most important necessities for life: food, shelter and clothing, focused on plants. Even though animal products were available, some form of clothing was needed that was lighter and cooler than skins and hides. It was easier to obtain from plants such items as bowstrings, nets, snares, etc. Also plant products were available from the leaves, stems and roots of many plants to construct shelter.

Very early on plant fibres have had a more extensive use than silk, wool and other animal fibres. Gradually as humans' needs multiplied, the use of vegetable fibres increased greatly until presently they continue to be of great importance even after the onset of plastics. It is impossible to estimate the number of species of fibre plants, but over a thousand species of plants have yielded fibres in America alone, and over 800 occur in the Philippines. However, plant fibres of commercial importance ore relatively few, the greater number being native species used locally by primitive peoples in all parts of the world. Their durability often exceeds those of synthetic manufacture, one example being sisal and Manila hemps. The most prominent fibres of the present are of great antiquity. The cultivation of flax, for example, dates back to the Stone Age of Europe, as discovered in the remains of the Swiss Lake Dwellers. Linen was used in Ancient Egypt and cotton was the ancient national textile of India, being used by all the aboriginal peop0les of the New World as well. Ramie or China grass has been grown in the Orient many thousands of years.

POLYMER FIBERS

Polymer fibers are a subset of man-made fibers, which are based on synthetic chemicals (often from petrochemical sources) rather than arising from natural materials by a purely physical process.

These fibers are made from:

- Polyamide nylon
- PET or PBT polyester
- Phenol-formaldehyde (PF)
- Polyvinyl chloride Fibre (PVC) vinyon
- Polyolefins (PP and PE) olefin Fibre
- Acrylic polyesters, pure polyester PAN fibers are used to make carbon Fibre by roasting them in a low oxygen environment. Traditional acrylic Fibre is used more often as a synthetic replacement for wool. Carbon fibers and PF fibers are noted as two resin-based fibers that are not thermoplastic, most others can be melted.
- Aromatic polyamids (aramids) such as Twaron, Kevlar and Nomex thermally degrade at high temperatures and do not melt. These fibers have strong bonding between Polymer chains
- Polyethylene (PE), eventually with extremely long chains/ HMPE (e.g. Dyneema or Spectra).
- Elastomers can even be used, e.g. spandex although urethane fibers are starting to replace spandex technology.
- Polyurethane Fibre
- Elastolefin

Coextruded fibers have two distinct polymers forming the Fibre, usually as a core-sheath or side-by-side. Coated fibers exist such as nickel-coated to provide static elimination, silver-coated to provide anti-bacterial properties and aluminum-coated to provide RF deflection for radar chaff. Radar chaff is actually a spool of continuous glass tow that has been aluminum coated. An aircraft-mounted high speed cutter chops it up as it spews from a moving aircraft to confuse radar signals.

ACRYLIC FIBRE

Acrylic fibers are synthetic fibers made from a polymer (polyacrylonitrile) with an average molecular weight of ~100,000, about 1900 monomer units. To be called acrylic in the U.S, the polymer must contain at least 85per cent acrylonitrile monomer. Typical comonomers are vinyl acetate or methyl acrylate. DuPont created the first acrylic fibers in 1941 and trademarked them under the name Orlon. Acrylic is also called acrilan fabric. It was first developed in the mid-1940s but was not produced in large quantities until the 1950s. Strong and warm, acrylic Fibre is often used for sweaters and tracksuits and as linings for boots and gloves, as well as in furnishing fabrics and carpets. It is manufactured as a filament, then cut into short staple lengths similar to wool hairs, and spun into yarn.Modacrylic is a modified acrylic Fibre that contains at least 35per cent and at most 85per cent acrylonitrile monomer. The comonomers vinyl chloride, vinylidene chloride or vinyl bromide used in modacrylic give the Fibre flame retardant properties. End-uses of modacrylic include faux fur, wigs, hair extensions and protective clothing.

Production

The polymer is formed by free-radical polymerization in aqueous suspension. The Fibre is produced by dissolving the polymer in a solvent such as N,N-dimethylformamide or aqueous sodium thiocyanate, metering it through a multi-hole spinnerette and coagulating the resultant filaments in an aqueous solution of the same solvent (wet spinning) or evaporating the solvent in a stream of heated inert gas (dry spinning). Washing, stretching, drying and crimping complete the processing. Acrylic fibers are produced in a range of deniers, typically from 0.9 to 15, as cut staple or as a 500,000 to 1 million filament tow. End uses include sweaters, hats, hand-knitting yarns, socks, rugs, awnings, boat covers, and upholstery; the Fibre is also used as "PAN" precursor for carbon Fibre. Production of acrylic fibers is centered in the Far East, Turkey, India, Mexico, and South America, though a number of European producers still continue to operate, including Dralon and Fisipe. US producers have ended production, though acrylic tow and staple are still spun into yarns in the USA. Former U.S. brands of acrylic were Acrilan (Monsanto), Creslan (American Cyanamid), and Orlon (DuPont). Other brand names that are still in use include Dralon (Dralon GmbH).

Textile uses

Acrylic is lightweight, soft, and warm, with a wool-like feel. It can also be made to mimic other fibers, such as cotton, when spun on short staple equipment. Some acrylic is extruded in colored or pigmented form; other is extruded in "ecru", otherwise known as "natural," "raw white," or "undyed." Pigmented Fibre has highest light-fastness. Its fibers are very resilient compared to both other synthetics and natural fibers. Some acrylic is used in clothing as a less expensive alternative to cashmere, due to the similar feeling of the materials. Some acrylic fabrics may fuzz or pill easily. Other fibers and fabrics are designed to minimize pilling. Acrylic takes Colour well, is washable, and is generally hypoallergenic. End-uses include socks, hats, gloves, scarves, sweaters, home furnishing fabrics, and awnings.Acrylic is resistant to moths, oils, chemicals, and is very resistant to deterioration from sunlight exposure. Acrylic is the "workhorse" hand-crafting Fibre for crafters who knit or crochet; acrylic yarn may be perceived as "cheap" because it is typically priced lower than its natural-Fibre counterparts, and because it lacks some of their properties, including softness and the ability to felt or take acid dyes. The Fibre requires heat to "kill" or set the shape of the finished garment, and it isn't as warm as alternatives like wool. Some knitters also complain that the Fibre "squeaks" when knitted, or that it is painful to knit with because of a lack of "give" or stretch in the yarn. On the other hand, it can be useful in certain items, like garments for babies, which require constant washing, because it is machine-washable and extremely Colour-fast. Acrylic can irritate the skin of people with dermatological conditions such as eczema.

ARAMID

Aramid fibers are a class of heat-resistant and strong synthetic fibers. They are used in aerospace and military applications, for ballistic rated body armor fabric and ballistic composites, in bicycle tires, and as an asbestos substitute. The name is a portmanteau of "aromatic polyamide". They are fibers in which the chain molecules are highly oriented along the Fibre axis, so the strength of the chemical bond can be exploited. Aromatic polyamides were first introduced in commercial applications in the early 1960s, with a meta-aramid Fibre produced by DuPont as HT-1 and then under the trade name Nomex.

This Fibre, which handles similarly to normal textile apparel fibers, is characterized by its excellent resistance to heat, as it neither melts nor ignites in normal levels of oxygen. It is used extensively in the production of protective apparel, air filtration, thermal and electrical insulation as well as a substitute for asbestos. Meta-aramid is also produced in the Netherlands and Japan by Teijin under the trade name Conex, In Korea by Toray under the trade name Arawin, in China by Yantai Tayho under the trade name New Star, by SRO Group (China) under the trade name X-Fiper, and a variant of meta-aramid in France by Kermel under the trade name Kermel. Based on earlier research by Monsanto Company and Bayer, a Fibre – para-aramid – with much higher tenacity and elastic modulus was also developed in the 1960s–1970s by DuPont and Akzo Nobel, both profiting from their knowledge of rayon, polyester and nylon processing. Much work was done by Stephanie Kwolek in 1961 while working at DuPont, and that company was the first to introduce a para-aramid called Kevlar in 1973. A similar Fibre called Twaron with roughly the same chemical structure was introduced by Akzo in 1978. Due to earlier patents on the production process, Akzo and DuPont engaged in a patent dispute in the 1980s. Twaron is currently owned by the Teijin company. Para-aramids are used in many high-tech applications, such as aerospace and military applications, for "bullet-proof" body armor fabric.

The Federal Trade Commission definition for aramid Fibre is:

- A manufactured Fibre in which the Fibre-forming substance is a long-chain synthetic polyamide in which at least 85per cent of the amide linkages, (-CO-NH-) are attached directly to two aromatic rings.

Health

During the 1990s, an in vitro test of aramid fibers showed they exhibited "many of the same effects on epithelial cells as did asbestos, including increased radiolabeled nucleotide incorporation into DNA and induction of ODC (ornithine decarboxylase) enzyme activity", raising the possibility of carcinogenic implications. However, in 2009, it was shown that inhaled aramid fibrils are shortened and quickly cleared from the body and pose little risk.

Production

World capacity of para-aramid production is estimated at about 41,000 tonnes/year in 2002 and increases each year by 5–10per cent. In 2007 this means a total production capacity of around 55,000 tonnes/year. Polymer preparation Aramids are generally prepared by the reaction between an amine group and a carboxylic acid halide group. Simple AB homopolymers may look like:

$nNH_{2-}Ar\text{-}COCl$! -(NH-Ar-CO)n- + nHCl

The most well-known aramids (Kevlar, Twaron, Nomex, New Star and Teijinconex) are AABB polymers. Nomex, Teijinconex and New Star contain predominantly the meta-linkage and are poly-metaphenylene isophthalamides (MPIA). Kevlar and Twaron are both p-phenylene terephthalamides (PPTA), the simplest form of the AABB para-polyaramide. PPTA is a product of p-phenylene diamine (PPD) and terephthaloyl dichloride (TDC or TCl). Production of PPTA relies on a co-solvent with an ionic component (calcium chloride ($CaCl_2$)) to occupy the hydrogen bonds of the amide groups, and an organic component (N-methyl pyrrolidone (NMP)) to dissolve the aromatic polymer. Prior to the invention of this process by Leo Vollbracht, who worked at the Dutch chemical firm Akzo, no practical means of dissolving the polymer was known. The use of this system led to an extended patent dispute between Akzo and DuPont. After production of the polymer, the aramid Fibre is produced by spinning the dissolved polymer to a solid Fibre from a liquid chemical blend. Polymer solvent for spinning PPTA is generally 100per cent anhydrous sulfuric acid (H_2SO_4).

Appearances

- Fibre
- Chopped Fibre
- Powder
- Pulp

OTHER TYPES OF ARAMIDS

Besides meta-aramids like Nomex, other variations belong to the aramid Fibre range. These are mainly of the copolyamide type, best known under the brand name Technora, as developed by Teijin and introduced in 1976. The manufacturing process of Technora reacts PPD and 3,4'-diaminodiphenylether

(3,4'-ODA) with terephthaloyl chloride (TCl). This relatively simple process uses only one amide solvent and therefore spinning can be done directly after the polymer production.

ARAMID FIBRE CHARACTERISTICS

Aramids share a high degree of orientation with other fibers such as ultra high molecular weight polyethylene, a characteristic that dominates their properties.

General:

- Good resistance to abrasion
- Good resistance to organic solvents
- Non-conductive
- No melting point, degradation starts from 500°C
- Low flammability
- Good fabric integrity at elevated temperatures
- Sensitive to acids and salts
- Sensitive to ultraviolet radiation
- Prone to static build-up unless finished

Para-aramids:

- Para-aramid fibers such as Kevlar and Twaron, provide outstanding strength-to-weight properties
- High Young's modulus
- High tenacity
- Low creep
- Low elongation at break (~3.5per cent)
- Difficult to dye – usually solution dyed

Major industrial uses:

- Flame-resistant clothing (example military MIL-G-181188B suits).
- Heat protective clothing and helmets
- Body armor, competing with PE based Fibre products such as Dyneema and Spectra
- Composite materials
- Asbestos replacement (e.g. brake linings)
- Hot air filtration fabrics
- Tires, newly as Sulfron (sulfur modified Twaron)
- Mechanical rubber goods reinforcement
- Ropes and cables
- Wicks for fire dancing
- Optical Fibre cable systems
- Sail cloth (not necessarily racing boat sails)
- Sporting goods
- Drumheads

- Wind instrument reeds, such as the Fibracell brand
- Loudspeaker diaphragms
- Boathull material
- Fibre reinforced concrete
- Reinforced thermoplastic pipes
- Tennis strings (e.g. by Ashaway and Prince tennis companies)
- Hockey sticks (normally in composition with such materials as wood and carbon)
- Snowboards
- Jet engine enclosures

TWARON

Twaron is the brandname of Teijin Aramid for a para-aramid. It is a heat-resistant and strong synthetic fibre developed in the early 1970s by the Dutch company AKZO, division ENKA, later Akzo Industrial Fibers. The research name of the para-aramid fibre was originally Fibre X, but it was soon called Arenka. Although the Dutch para-aramid Fibre was developed only a little later than DuPont's Kevlar, introduction of Twaron as a commercial product came much later than Kevlar due to financial problems at the AKZO company in the 1970s.

This is a chronology of the development of Twaron:

- In 1960s a research Programme starts for "Fibre X."
- In 1972 the ENKA Research laboratory develops a para-aramid called Arenka.
- In 1973 Akzo decides to use sulfuric acid (H_2SO_4) as a solvent for spinning.
- In 1976 a pilot plant is built.
- In 1986 commercial production is started at five locations and nine plants.
- In 1987 Twaron is introduced as a commercial product.
- In 1989 the aramid business of Akzo becomes an independent Business Unit called Twaron BV.
- Since 2000 Twaron BV is owned by the Teijin Group, now called Teijin Twaron BV and based in Arnhem, The Netherlands. The main production facilities for Twaron are in Emmen and Delfzijl.
- In 2007 Teijin Twaron expands for the fourth time in six years and also changes its name into Teijin Aramid.

PRODUCTION

Polymer Preparation

Twaron is a p-phenylene terephthalamide (PpPTA), the simplest form of the AABB para polyaramide. PpPTA is a product of p-phenylene diamine (PPD) and terephthaloyl dichloride (TDC). To dissolve the aromatic polymer Twaron used a co-solvent of N-methyl pyrrolidone (NMP) and an ionic component (calcium chloride $CaCl_2$) to occupy the hydrogen bonds of the amide groups. Prior to the invention of this process by Leo Vollbracht, working at the Dutch chemical firm AKZO, no practical means of dissolving the polymer was known. The use of this system by DuPont led to a patent war between AKZO and DuPont as Dupont initially used the carcinogenic HMPT (hexamethylphosphoramide). Despite heavy research DuPont now also uses the AKZO patent to use the less hazardous NMP in the Kevlar process.

Spinning

After the production of the Twaron polymer in Delfzijl, the polymer is brought to Emmen, where fibres are produced by spinning the dissolved polymer into a solid fibre from a liquid chemical blend. Polymer solvent for spinning PPTA is generally 100per cent anhydrous (water free) sulfuric acid (H_2SO_4).The polymer is dissolved by mixing frozen sulfuric acid in powder form with the polymer in powder form and gently heating the mixture. This process, which differs from the more difficult DuPont process, was invented by Henri Lammers and patented by AKZO.

INDUSTRIAL USES

Twaron is a para-aramid and is used in automotive, construction, sport, aerospace, military and industry applications, e.g., "bullet-proof" body armor, fabric, and as an asbestos substitute.

- Protective gear (heat resistant/ ballistics): flame-resistant clothing, protective clothing and helmets, cut-fast or heat-hardy gloves, sporting goods, textiles, ballistic vests
- Composites: composite materials, technical paper, asbestos replacement, hot air filtration, sailcloth, speaker woofers, boat hull material, Fibre reinforced concrete, drumheads
- Automotive: brake pads, turbo hoses, V-belts and Timing belts, tires that incorporate Sulfron (sulfur modified Twaron), mechanical rubber goods reinforcement
- Linear tension: optical Fibre cables (OFC), ropes, wire ropes, cables, umbilical cables, electrical mechanical cable (EMC), reinforced thermoplastic pipes

KEVLAR

Kevlar is the registered trademark for a para-aramid synthetic Fibre, related to other aramids such as Nomex and Technora. Developed by Stephanie Kwolek at DuPont in 1965, this high-strength material was first commercially used in the early 1970s as a replacement for steel in racing tires. Typically it is spun into ropes or fabric sheets that can be used as such or as an ingredient in composite material components.Currently, Kevlar has many applications, ranging from bicycle tires and racing sails to body armor because of its high tensile strength-to-weight ratio; by this measure it is 5 times stronger than steel. It is also used to make modern drumheads that withstand high impact. When used as a woven material, it is suitable for mooring lines and other underwater applications. A similar Fibre called Twaron with roughly the same chemical structure was developed by Akzo in the 1970s; commercial production started in 1986, and Twaron is now manufactured by Teijin. Poly-paraphenylene terephthalamide – branded Kevlar – was invented by Polish-American chemist Stephanie Kwolek while working for DuPont, in anticipation of a gasoline shortage. In 1964, her group began searching for a new lightweight strong Fibre to use for light but strong tires. The polymers she had been working with at the time, poly-p-Phenylene-terephthalate and polybenzamide, formed liquid crystal while in solution, something unique to those polymers at the time. The solution was "cloudy, opalescent upon being stirred, and of low viscosity" and usually was thrown away. However, Kwolek persuaded the technician, Charles Smullen, who ran the "spinneret", to test her solution, and was amazed to find that the Fibre did not break, unlike nylon. Her supervisor and her laboratory director understood the significance of her accidental discovery and a new field of polymer chemistry quickly arose. By 1971, modern Kevlar was introduced. However, Kwolek was not very involved in developing the applications of Kevlar.

ROUGH WEAVING & PLAITING FIBERS

There are relatively few materials that are manufactured for plaited or coarsely woven articles. The raw materials include the rushes, stems of reeds, willows, bamboo, grasses, rattan and leaves and roots. They are used entirely or split. They are woven or twisted together in a simple manner and made into sandals, mats, hats, matting, screens, chair seats, baskets, etc.

HAT FIBERS

In many parts of the Eastern Hemisphere, rice, barley, wheat and rye are grown for the purpose of making braids or straw plaits for hats. The plants are grown close together so that they will have few leaves, and they are harvested before they mature. The stems are split lengthwise before plaiting. The Leghorn Hats and Tuscan Hats of Italy are some of the best of the straw hats. Panama Hats are made from the leaves of Toquilla, Carludovica palmata, a stem less, palm like plant that grows wild in the forests from southern Mexico to Peru. It has been cultivated in Ecuador and parts of Colombia. The Panama hat industry is concentrated in Ecuador. Young leaves are collected while they are still folded in the bud and treated with hot water. The coarse veins are removed and the plaits are separated and split lengthwise into slender strips that are slowly dried and bleached. They gradually roll inward forming fine cylindrical strands known as Jipijapa. The hats are woven by hand from these strands. About six leaves are necessary to make one hat. The best quality Panama hats are uniform and have a fine texture, are strong, durable and elastic and resistant to water. The Puerto Rican Hats are made from the leaves of the hat palm,Sabal causiarum.

MATS AND MATTING

In the Eastern Hemisphere commercial mattings have been made from several rushes, grasses and sedges. Usually the stalks or leaves are used alone, but they may be combined with cotton of hemp. Some of the species utilized are Chinese Mat Grass, Cyperus tegetiformis, and Japanese Mat Rush, Juncus effusus. The Screw Pines, Pandanus tectorius and P. utilis are important in Southeastern Asia and Oceania for making mats. The leaves of these species are also used for sugar bags, cordage, hats and thatching.

BASKETS

Baskets have been and are continuously being made from an array of plant species worldwide. Roots, stems, leaves and even woody splints have been used. Commercial baskets are usually made from rushes, cereal straw, osiers or willows, and ash or white oak splints. Sweet grass baskets are made from Hierochloe odorata, a common species in lowlands along the coast and Great Lakes. Another important source of basket fiber is the raffia palm, Raffia pedunculata, native to Madagascar. Strips of the lower epidermis of the leaves are the raffia of commerce. The fiber is so soft and silk like that it can be woven. It is especially useful as a tie material for nurseries and gardens.

WICKERWORK

This includes chair seats, chairs, infant carriages, hampers and other light articles of furniture. Willows, rattan and bamboo are the main plants used. Rattan is obtained from several species of climbing palms, Calamus spp., that grow in the humid forests of the East Indies and other parts of tropical Asia. The stems of these plants are long, strong, flexible and uniform. They are used either entirely or as splits in Asia for furniture, canes, baskets and other items. A considerable quantity of rattan is exported for making furniture. Bamboos occur in most tropical areas, but they are especially abundant in the monsoon regions of Eastern Asia. They are the largest of the grasses with woody stems that sometimes reach one foot in diameter and a height of over 10 feet. There are many species in the families Arundinaria, Bambusa, Dendrocalamus, Gigantochloa, Phyllostachys, and other genera. The stems are used for all kinds of construction in areas where these plants grow. Exported bamboo is used in the manufacture of furniture, fishing rods and implements of various kinds. Bamboo splits are made into baskets and brushes. In the Western

Hemisphere bamboos have not been extensively utilized. Guadua angustifolia is a species with very strong culms and has been used in Ecuador to make furniture and in house construction.

FILLING FIBERS

Many plant fibers have been used to stuff pillows, cushions, furniture, mattresses, etc. They are also used to caulk the seams of vessels, in the making of staff for buildings, as stiffening for plaster, packing for bulkheads and machine bearings, and for the protection of delicate objects during shipment. Synthetic materials frequently take the place of these long used products, but in some ways they retain some superiority. Surface fibers are commonly used for stuffing because their staples are too short to be spun and thus are not valued in the textile industry. Bast fibers are too costly, and hard fibers are frequently too stiff and coarse. The silk cottons are the most important source for stuffing.

KAPOK

This is the most popular silk cotton and most valuable of all the stuffing substances. Kapok is the floss produced in the pods of the kapok tree, Ceiba pentandra. Originally confined to the American tropics, it is now found worldwide. It is an irregular tree, 50-100 ft. tall, with a buttressed base and weird growth habit. It grows rapidly and begins to bare when only 15 ft. tall. A mature tree can produce more than 600 pods and from 6-10 lbs. of the cottony fibers. Pods are clipped from the branches and opened. The floss is removed and the seeds separated by centrifugal force. The floss is 1/2-1 1/2 in. long and whitish, yellowish or brownish in color. It is very fluffy, light and elastic and is thus an ideal stuffing material for mattresses and pillows. The fibers have a low specific gravity. They are five times more buoyant than cork and are impervious to water. Therefore, kapok is valuable as a filling for life preservers, cushions, portable pontoons, etc. Its low thermal conductivity and its high ability to absorb sound make kapok an excellent material for insulating small refrigerators and for soundproofing rooms. It has also been used for the linings of sleeping bags, gloves for handling dry ice and in the tropics as surgical dressings. Kapok seeds have 45 percent fatty oil that is extracted and used for soap and food.

KAPOK SUBSTITUTES

There are a number of other plants with seed hairs or floss that can be used as a substitute for kapok. The Red Silk Cotton Tree or Simal, Salmalia malabarica, is a very large ornamental tree. It supplies reddish floss known as Indian Kapok that has been important as a stuffing in India for centuries. The White Silk Cotton Tree, Cochlospermum religiosum, yields a fiber of some importance. This handsome tree is native to India but is now widespread in the world tropics. It is also one source of Kadaya Gum. Madar, Calotropis gigantea, and the related Akund, Calotropis procera, are shrubs native to Southern Asia and Africa that produce a silk of some importance. Although inferior to kapok, this substance is often used in mixtures with kapok. The Pochotes of Mexico, Ceiba aesculifolia, C. acuminata, etc., yield a silk cotton almost equal to kapok in buoyancy and resiliency. Palo Borracho, Chorisia insignis, and Samohu, Chorisia speciosa, of South America yield large amounts of a glossy, white silk cotton with properties similar to kapok. All of the milkweeds have silky hairs on their seeds and several species are a source of stuffing materials. Milkweed floss is one of the lightest materials. it is very buoyant and a perfect insulator. It was used during World War II as a substitute for kapok. The pods contain oil and a wax that may have future applications. Some species yield textile fibers. In North America, Asclepias syriaca and A. incarnata produce abundant floss. In the Neotropics, A. curassavica has some value.

MISC. FILLING FIBERS

There are innumerable plants and fibers that have use as filling materials. Included are cereal straw, cornhusks, Spanish Moss and Crin Végétal. Spanish Moss, Tillandsia usneoides, is a conspicuous tree epiphyte in Southeastern North America. This is an excellent substitute for horsehair after it is processed. The plant is pulled from the trees with rakes or hooks, or it is collected from the ground or water. It is then fermented in order to rot off the gray outer covering and ginned to remove impurities. The prepared fiber is brown or black, lustrous and very resilient. It has been used in upholstery and for automobile cushions. Crin Végétal. Chamaerop humilis a dwarf fan palm of Northern Africa and the Mediterranean region in which the leaves have shredded and twisted fibers. These have been used as stuffing material.

NATURAL FABRICS

Some trees have basts with tough interlacing fibers that can be extracted from the bark in layers or sheets and can then be pounded into rough substitutes for cloth. Tapa Cloth is one of these as it once constituted the main clothing in Polynesia and parts of Eastern Asia. The material is obtained from the bark of the paper mulberry, Broussonetia papyrifera. Strips of bark are peeled from the trunk and the outer coating is scraped away. After soaking in water and cleaning these strips are placed on a hardwood log and pounded with a mallet. Overlapping the edges and beating them together unite the individual strips. The finished product varies according to thickness from muslin like material to one of leather. Tapa cloth is frequently dyed. Similar bark cloths have been manufactured from different sources since antiquity. In South America the Amerindians used the Tauary, Couratari tauari, and other species of the same genus. In Mozambique the wild fig, Ficus nekbudu, was used as a source of Mutshu Cloth. The Upas Tree, Antiaris toxicaria, of Sri Lanka furnishes a bark cloth. it is also the source of an important poison used with arrows. Lace Bark is the produce of Lagetta lintearia, a small tree of Jamaica. The inner bark is removed in sheets and can be stretched into a lacelike material with pentagonal meshes. It is suitable as a textile and ornament. Cuba Bast is from Hibiscus elatus, a small bushy tree of the West Indies. The inner bark is removed in long ribbon-like strips that have been used in millinery and for tying cigars. The vegetable sponges, Luffa cylindrica and L. acutangula, yield a unique fiber. These are climbing cucumbers of the tropics that bear edible fruits containing a lacy network of stiff curled fibers. This material is extracted by retting in water. After cleaning it is used for making hats, for washing and scouring machinery, in certain types of oil filters and as a substitute for bath sponges. A large amount of this material used to be exported by Japan.

MANUFACTURE OF SYNTHETIC FIBRESMANUFACTURE OF SYNTHETIC FIBRES

Most synthetic fibres are made by forcing liquids through tiny holes in a metal plate and allowing them to harden. A wide range of liquids produces a great variety of fibres. The metal plates are called spinnerets. They are made of gold or platinum because these metals are not affected by most chemicals. The size of the spinneret is about the size of thimble and it has 10 to 150 small openings, depending on the thickness of the strand wanted. Different synthetic fibres are made from different raw materaials.

RAYON

It is also called artificial silk. Rayon is made from cellulose. There are several varieties of rayon. Buyt rayon produced by the viscose process is the most important.

The ingredients for making viscose rayon are:

1. Cellulose ($C_6H\text{-}10O_5$),
2. Sodium hydroxide (NaOH),
3. Carbon disulphide (CS_2), and
4. Sulphuric acid (H_2SO_4).

Manufacture of Rayon

The manufacture of rayon involves the following steps.

1. Cellulose in the form of wood pulp is treated with NaOH.
2. On adding CS_2, it dissolves completely and a yellow syrup-like liquid called viscose is formed.
3. Viscose is forced through the fine holes of the spinneret into a solution of dilute H_2SO_4. Silk-like threads are formed. This product is viscose

rayon.

Uses of Rayon

Rayon can be mixed with cotton or wool, which makes it more suitable for our needs. It is a good fabric or sarees. When mixed with cotton it makes a good dress material. Aprons and caps are preferably made of rayon. On mixing it with wool, it serves as a good fibre for making carpets. Bandages and lints for dressing wounds are made of rayon. Hosepipes and conveyor belts are also made from rayon. Perspiration weakens rayon fibres and they lose strength when wet.

Acetate

Acetate is another well-known fibre made from wood pulp. The reaction between wood pulp (cellulose) and acetic acid is the basis for this manufacture of this fibre. Acetate is made of fibres that do not wrinkle or shrink as much as rayon. It is an efficient smoke remover, thus it is used in cigarette filters. Acetate fibres melt when burned. They are destroyed by pressing with very hot iron. Some dry-cleaning solvents dissolve the fibres.

Nylon

Nylon is a polymer made of polyamide chains. The basic materials for making nylon are coal and petroleum. The polymer is squirted through spinneret holes to form nylon threads. The strands are then stretched four times their original length. The stretching forces the molecules to line up, giving nylon an increased strength and making it more elastic. Nylon is light weight, fine and durable. It is resistant to moths and molds. It absorbs very little water, therefore it dries quickly.

Uses of Nylon

Hammocks, fishing nets tyre cords, ropes, bristles of brushes and parachute fabrics are all made of nylon fibre. As nylon is elastic in nature, it is a good material for making stockings and socks. Nylon sarees are quite common in our country.

Nylon has a few weaknesses, as it absorbs very little moisture it is difficult to dye. It produces static electricity when rubbed. Being a non-cellulose fibre, it requires low to moderate ironing heat.

Polyester

Acrylic and polyester are non-cellulose fibres. They are manufactured from petroleum products. Terylene and Dacron are examples of polyesters. These fibres are easy to wash; they dry quickly, and resist chemicals and wrinkles. They are difficult to dye. These fibres blend well with natural fibres in making cloth. Terylene is often mixed with cotton to make terycot, with wool, it gives terywool. Clothes made of these are more comfortable to wear than pure terylene.

DYEING OF ENVIRONMENTALLY FRIENDLY PRETREATED COTTON FABRIC

Raw cotton fibres have to go through several chemical processes to obtain properties suitable for further dyeing and use. With scouring, non-cellulose substances (wax, pectin, proteins, hemicelluloses...) that surround the fibre cellulose core are removed, and as a result, fibres become hydrophilic.

Conventional scouring processes of cotton are conducted at temperatures up to 130 °C in a very alkaline medium with sodium hydroxide. Since a non-specific reagent is used in the treatment, it attacks impurities but it also causes damages to the cellulose portion of the fibres. Several auxiliary agents, such as wetting agents, emulsifiers and sequestering agents, which improve the efficiency of scouring and reduce the damage of fibres, are also added to the scouring bath.

Scouring is regularly followed by a bleaching process, which removes the natural pigments of cotton fibres. Cellulose fibres are most frequently bleached with hydrogen peroxide resulting in high and uniform degrees of whiteness. The water absorbency also increases, however, during the decomposition of hydrogen peroxide, radicals that can damage the fibres are formed. For this reason, organic and inorganic stabilisers and sequestering agents are added to the treatment bath. Hydrogen peroxide is not ecologically disputable.

The large amount of water used to rinse and neutralise the alkaline scoured and peroxide bleached textiles is ecologically disputable. Namely, the bleaching process is conducted in an alkaline bath at pH 10 to 12 and at temperatures up to 120 °C. Due to high working temperature, a large amount of energy is consumed.

Auxiliary chemicals added into the bath increase the TOC and COD values of effluents. Upon neutralisation of highly alkaline waste baths, large amounts of salts are produced. Consequently, the textile industry is considered one of the biggest water, energy and chemical consumers. To comply with more and more rigourous environmental regulations and to save water and energy, biotechnology and several types of enzymes have entered the textile sector.

Many review and scientific papers describe the use of different enzymes for textile finishing. Pectinases are an efficient alternative to sodium hydroxide in the removal of non-cellulose substances from the cotton fibre surface. This process occurs at moderate temperatures in a slightly acidic or alkaline medium and is dependent on the type of pectinase. Pectin acts as a sort of cement or matrix that stabilises the primary cell wall of cotton fibres. Pectinases decompose insoluble pectin into smaller particles thereby destabilising the structure in the outer layers. The weakened outer layers can be removed in a subsequent wash process to such extent that following finishing processes and dyeing can be easily preformed.

It has been established that agitation of the treatment bath is very important for pectinase to function efficiently and that selected sequestering agents can improve their effectiveness. Since enzymes act selectively, no damage to fibres occur during treatment. Also, it has been observed that after enzymatic treatment, baths are less polluted than baths after scouring with sodium hydroxide.

Bleaching with peracetic acid (PAA) is an alternative to bleaching with hydrogen peroxide. It is a powerful oxidising agent with excellent antimicrobial and bleaching properties. It is efficient at low concentrations, temperatures and in neutral to slightly alkaline medium. Its products of decomposition are biologically degradable. In the past, it was prepared in situ from acetic acid anhydride and hydrogen peroxide. However, the risk of explosion during the synthesis reaction prevented affirmation of peracetic acid as a bleaching agent in industry. In recent years, peracetic acid has become interesting. Several commercial products are available as balanced mixtures of peracetic acid, acetic acid and hydrogen peroxide. They are stabilised with a minimum amount of sequestering agent. Today, peracetic acid products available in the market are safe, simple to use, and priceeffective. Equation 2 shows the reaction that occurs when peracetic acid is used for bleaching.

Both processes, scouring with pectinases and bleaching with peracetic acid, are conducted at temperatures of 50–60 °C for 40–60 minutes and pH 5–8. If both processes could be combined into one process, huge amounts of water, energy, time, and auxiliary agents can be saved. In a recent study, it was confirmed using a viscosimetric method that pectinases retain their activity in the presence of peracetic acid and that combined processes are feasible.

The objective of our work was to compare the properties of enzymatically-scoured and peracetic acid-bleached cotton fabrics treated by two-bath and one-bath scouring/bleaching methods, with respect to conventionally-treated fabrics (alkaline scoured and bleached with hydrogen peroxide) with emphasis on their degree of whiteness and dyeability. Dyeability of differently pretreated fabrics is very important for industry and was studied from many aspects before.

The dyeing behaviour of cotton fabrics treated with different enzymes by using reactive, cationic and acid dyes showed that cationic and acid dyes were more sensitive to the enzymatic treatment used as reactive dyes. When dyed with bireactive dyes the dyeing showed excellent evenness and their capacity to cover differences in whiteness arising from different pretreatment processes was significant. On the contrary, the dyeings with acid and cationic dyes revealed, with large differences in dye exhaustions, that each enzymatic system produces different ionisable residues on the primary wall of the cotton fibre. The importance of charges and functional availability on the fibre surface after different bioscouring treatments for their dyeing behaviour was also exposed by Calafell and co-workers. The ability of reactive dyes to cover the differences in whiteness from different bioscouring and bleaching processes when dyed in dark shades was confirmed by several researchers. Our research will upgrade the above results with reactive dyeing of differently scoured and differently bleached cotton fibres in medium and light shades and estimate the processes from the ecological point of view.

MATERIALS

Desised cotton fabric, 100 g/m^2, was obtained from Tekstina, Slovenia. Acid pectinase Forylase KL (AP) was supplied from Cognis, Germany, and alkaline pectinase Bioprep 3000L (BP) from Novozymes, Denmark. Cotoblanc HTD-N (anionic wetting and dispersing agent, alkansulphonate with chelator) was supplied from CHT, Germany. H_2O_2 35 per cent (HP) and peracetic acid (PAA) as a 15 per cent equilibrium solution in the commercial bleaching agent Persan S15 were obtained from Belinka, Slovenia. Foryl JA (nonionic wetting agent) and Locanit S (ionic-nonionic dispersing agent) were obtained from Cognis, Germany and Lawotan RWS (nonionic wetting agent) was obtained from CHT, Germany. Sodium hydroxide was supplied from Šampionka, Slovenia, and acetic acid and sodium carbonate were supplied from Riedel-de Haen, Germany.

TREATMENT METHODS

The cotton fabric was scoured according to three different procedures using sodium hydroxide, acid pectinase or alkaline pectinase. The scoured fabrics were bleached with two bleaching agents: hydrogen peroxide and Persan S15. The abbreviation of processes and treatment conditions are displayed in table.

Table. The Abbreviation of Processes and Treatment Conditions.

Process	Conditions
AS-Alkaline Scouring	3 g/l NaOH, 2 g/l Cotoblanc HTD-N, 95 °C, 40 minutes
AP-Scouring with Acid Pectinases	5ml/1 Forylase KL, 0.75ml/1Fory1 2ml/1 Locanit S and CH_3COOH pH 5.5, 55°C, 40 min.
BP-Scouring with alkaline pectinases	0.05% Bioprep, 0.5 g/1 Lawotan RWS, Na_2CO_3 to pH 8,55°C, 40 min
HP-Bleaching with hydrogen peroxide.	7g/1 H_2O_2 35%, 1 g/1 Cottoblanc HTD-N 4 g/1 NaOH 100%, 95°C, 40 min
PAA- Bleaching with peracetic acid	15 ml/1 Persan S15, 55 ml/1 Na_2CO_3 0.5 M, 0.1g/1 Lawotan RWS, pH 8, 55°C, 40 min
AP+PAA- one step scouting	5 ml/1 Forylase KL, 0.75 ml/1

with and pectinase and bleaching with peracetic acid	fory1 JA, 2ml/1 LocanitS, 15 ml/1 Persan S15
BP+PAA-one step scouring pectinase and bleaching with peracetic acid	0.05% Bioprep 3000L, 0.1 mL/L with alkaline Lawotan RWS, 15 mL,/L Persan S15, pH 8 with NaOH, 55° C, 40 min.

Enzymatic scouring and one-step treatments were performed 60 minutes at 55 °C, than the temperature of the bath was increased to 80 °C for 10 minutes to deactivate the enzymes. To activate peracetic acid in AP/PAA treatment, the pH was adjusted to 8 after 30 minutes. Demineralised water was used in all processes. The treatments were performed on the Jet JFL apparatus manufactured by Werner Mathis AG loaded with 50 g of fabric at a liquor ratio of 1:20. After all treatments, the bath was discharged and the jet was filled sequentially with fresh water heated to 80 °C, 60 °C and 25 °C to rinse the fabric. After alkaline scouring and peroxide bleaching, the fabrics were additionally neutralised with a neutralising bath containing acetic acid and rinsed with cold water.

DYEING PROCEDURE

Dyeing the pre-treated fabrics was performed at 60 °C for 90 minutes with 0.5 per cent and 2 per cent Cibacron rot F-B. 30 g/L Na2SO4 and 8 g/L Na_2CO_3 was used for pale shade and 50 g/L Na_2SO_4 and 11 g/L Na_2CO_3 for medium shade. The weight of the dyed samples was 5 g. Finally, the cotton was soaped, washed and air dried. Dyeing was performed in closed beakers on an Launder-Ometer (Atlas).

ANALYTICAL METHODS

Prior to the measurements, fabrics were conditioned 24 hours at 20 °C and 65 per cent relative humidity. The degree of whiteness and the colour values were measured on the Spectraflash SF600 Plus using the CIE method according to EN ISO 105-J02:1997(E) standard and EN ISO 105-J01:1997(E), respectively. Weight loss due to the pretreatments was determined by weighing the fabric samples before and after pretreatment and was expressed in percent.

Water absorbency was measured according to DIN 53 924 (velocity of soaking water of textile fabrics, method for determining the rising height). Measurements of tenacity at maximum load were performed on Instron Tensile Tester Model 5567.

The mean degree of polymerisation (DP) was determined with the viscosimetric method in cuoxam. Samples of remaining bleaching and scouring baths were collected after all treatments. Their ecological parameters (pH, total organic carbon (TOC), chemical oxygen demand (COD), biological oxygen demand (BOD5)) were measured, the consumption of water and energy was estimated.

WHITENESS

The achieved degrees of whiteness (W) and tint values (TV) are presented in table.

The desised (untreated) sample (D) had a degree of whiteness of 11.1. After alkaline scouring, the fibres swelled, became smoother and clean of non-cellulose impurities and the degree of whiteness increased to 19.5.

Table. Whiteness (W) and tint Values (TV).

Scouring	HP	PAA	Scouring/PAA									
	W	TV	W	TV	W	TV	W	TV	D	11.1	-9.1	- -
-	-	-	-	AS	19.5	-7.6	84.12	-0.45	72.7	-0.95	-	- AP
8.2	-9.73	85.59	-0.37	57.7	-2.05	68.7	-1.3	BP	8.4	-9.6	85.07	-0.36
57.3	-1.9	69.6	-1.2									

However, after scouring with acidic and alkaline pectinases, both samples had lower degrees of whiteness relative to the desised sample, *i.e.,* the sample treated with acidic pectinases (AP) had a whiteness degree of 8.2, and the sample treated with alkaline pectinases (BP) had a whiteness degree of 8.4. Negative TV values demonstrated that all scoured and desised samples had a red shade. After alkaline scouring, the red shade decreased, whereas after both bioscourings, the red shade increased. The degree of whiteness of all scoured samples increased significantly after hydrogen peroxide bleaching.

The differences in whiteness from previous scouring disappeared. Alkaline and bioscoured samples have a whiteness values above 84 and the red shade almost disappeared. With peracetic acid bleaching, a high degree of whiteness was not achieved and the differences in whiteness from the previous scouring remained visible. The sample, which was alkaline scoured prior to bleaching (AS+PAA), had the highest degree of whiteness (72.7), whereas both bioscoured samples had lower degrees of whiteness (57.7 AP+PAA and 57.3 AP+PAA). The red shade was visible on all peracetic acid bleached samples and was more on bioscoured than on alkaline scoured fabrics, which suggests that bleaching with peracetic acid is not as effective as bleaching with hydrogen peroxide. This occurs because bleaching with peracetic acid proceeds at a low temperature and pH, where the impurities remaining after scouring could not be fully oxidised. Bioscoured fibres contained also more waxes and other impurities that hindered the successful oxidation with peracetic acid at mild conditions. Bleaching the alkaline scoured fabrics with peracetic acid is more effective since the impurities were removed from cotton fibres to a higher extent in the previous process and the pigments within fibres were more exposed to the oxidant's influence. This is confirmed by comparing data of the mass loss during treatments.

Table. The Loss of Mass. Rising Height in Warp Direction Tenacity at Maximum Load, Degree.

	Weight loss (%)	Rising Height (cm)	Tenacity (cN/tex)	DP
D		0	18.47	2482
AS	1.27	2.9	18.45	2432
AP	0.30	2.7	16.96	2451
BP	0.89	2.5	17.95	2385
AS+HP	1.52	3.0	16.65	1774
AP+HP	1.51	3.0	17.12	1947
BP+HP	1.62	2.8	16.83	2004
AS+PAA	1.30	2.8	16.94	2278
AP+PAA	0.65	2.9	18.12	2318
BP+PAA	0.95	2.9	13.75	2399
AP/PAA	0.40	2.7	16.94	2438
BP/PAA	0.60	2.8	18.84	2300

The degrees of whiteness after a one-bath treatment were higher than those after two-bath bioscouring and bleaching with peracetic acid and close to the whiteness achieved after alkaline scouring and bleaching with peracetic acid.

This can be explained by the fact that at the bleaching conditions of peracetic acid, hydrogen peroxide, which was present in the balanced mixture with peracetic acid, was not consumed, whereas at temperature 80 °C, which was the finite temperature of the one-bath process, hydrogen peroxide was activated, which further increased the degree of whiteness.

FABRIC PROPERTIES

Table represents the loss of weight, rising height in warp direction, tenacity at maximum load and degree of polymerisation (DP) of differently pretreated cotton fabric samples.

The loss of weight demonstrates that scouring with NaOH is more intensive and removes more incrusts than enzymatic scouring. The loss of mass after alkaline scouring was ca. 1.3 per cent and after enzymatic scouring, was less than 1 per cent.

During hydrogen peroxide bleaching, the loss of mass was greater in those samples, where the loss of mass was lower during scouring; alkaline scoured only 0.25 per cent, acid pectinases scoured 1.2 per cent and alkaline pectinases scoured 0.73 per cent. This suggests that hydrogen peroxide bleaching removed a large portion of compounds, which remained on fibres after scouring. The total mass loss after scouring and hydrogen peroxide bleaching was similar for all samples.

Peracetic acid bleaching also removed a certain part of the noncellulosic substances, which remained on fibres after scouring, but the quantity was lower relative to hydrogen peroxide bleaching. Bleaching with peracetic acid did not equalise the differences in the loss of mass, which is in agreement with the whiteness results.

We can conclude that high temperature and high pH are conditions that contribute decisively to the removal of non-cellulosic impurities. Specifically, waxes cannot be removed completely when all processes are conducted at low temperatures and neutral pH, as is the case for bioscouring and peracetic acid bleaching.

The remained substances influence on the water absorbency and consequently alkaline scoured samples had the highest absorbency. Bleaching improved the absorbency of the scoured fabrics, particularly of enzymatically scoured ones. However, the difference in rising height was so small, that all the samples could be considered absorbent. There were no higher differences in tenacity at maximum load between the de-sized and differently treated samples. On the other hand, the results of DP demonstrate that bleaching with hydrogen peroxide decreased the degree of polymerisation significantly, while other processes preserved the DP values close to the starting value. The bioscouring and bleaching with peracetic acid in a one bath or two bath processes causes no damage to fibres and this is one of the benefits of such processes.

DYEING

Table. Colour Strengths (K/S) and Colour Differences (DE*) of the Fabric Samples Dyed with 0.5 Per cent of Cibacron Red F-B.

Scouring	HP	PAA	Scouring/PAA						
	K/S	Δ E	K/S	Δ E*	K/S	Δ E*	K/S	Δ E*	D
4.02	-	-	-	-	-	-	-	AS	4.04
-a	2.61	-b	3.89	-c	-	-	AP	4.04	3.74
2.73	0.26	3.89	1.26	3.78	0.58	BP	4.23	3.04	2.72
0.22	4.10	0.85	3.97	0.30					

Note:

a-standard for scoured fabrics, b-standard for HP bleached samples, c-standard for PAA bleached andone-bath treated fabrics

Table. Lightness Values (L*), Croma (C*) and hue (h) of the Fabric Samples Dyed with 0.5%.

Scouring	HP	PAA	Scouring/PAA					
	L*	C*	h [°]	L*	C*	h [°]	L*	C*
h [°]	L*	C* h [°]						
D	57.4	44.4	351.8	-	-	-	-	-
-	-	-	-					
AS	58.4	46.1	351.0	59.5	48.8	350.8	59.3	47.4
350.5	-	-	-					
AP	57.4	43.6	351.6	59.0	49.4	350.7	58.9	46.2
350.6	59.3	46.9 350.2						
BP	57.7	43.9	351.8	59.1	49.3	350.9	59.1	46.6
350.4	59.4	47.2 350.3						

Table. Colour Strengths (K/S) and Colour Differences (˜E*) of the Fabric Samples Dyed with 2% of Cibacron Red F-B and Standard Deviation in Brackets.

Scouring				HP	PAA		Scouring/PAA				
	K/S	Δ E*	K/S	Δ E*	K/S	Δ E*	K/S	Δ E*	D	13.24	-
-	-	-	-	-	-	AS	13.42	-a	9.72	-b	
13.38	-c	-	-	AP	14.17	0.84	9.57	0.03	13.10	1.15	
13.190.81		BP	13.33	1.27	9.55	0.03	13.26	1.00	14.25	0.65	

Note:

a-standard for scoured fabrics, b-standard for HP bleached samples, c-standard for PAA bleached andone-bath treated fabrics

Table. Lightness Values (L*) Croma (C*) and hue (h) of the Fabric Samples Dyed with 2 % of Cibacron Red F-B and Standard Deviation in Brackets.

Scouring	HP	PAA	Scouring/PAA								
	L*	C*	h [°]	L*	C*	h [°]	L*	C*	h [°]	L*	C* h [°]
D	44.7	54.6	356.0	-	-	-	-	-	-	-	- -
AS	45.3	56.0	356.2	46.0	58.5	357.0	46.2	57.0	356.1	-	- -
AP	44.6	54.9	356.3	46.2	58.5	356.9	45.7	56.0	355.8	45.7	56.4 355.9
BP	44.1	55.0	356.4	58.4	356.9	45.4	56.4	356.0	45.7	56.6	56.6 356.0

Colour strengths (expressed as K/S value) and colour differences, ΔE*, of the fabric samples dyed with 0.5 per cent and 2 per cent of Cibacron red F-B, respectively. The lightness values, L*, croma, C*, and hue, h, are presented in tables. A standard for colour difference calculation was in each set of processes the alkaline scoured sample. K/S values of all samples dyed with one concentration of dye after only scouring are similar. No significant differences were observed between the colour depths of alkaline and enzymatic pretreatments.

On the contrary, colour differences between alkaline scoured and enzymatic scoured samples dyed in light shades are significant. This is explained by differences in whiteness after scouring, which was not covered in light shade dyeing, whereas they were covered in dark shade dyeing to a greater extent. The colour strengths of the samples dyed after scouring and bleaching are very similar within a set of samples, but they differ between differently bleached samples. K/S values on all hydrogen peroxide bleached samples were lower than those obtained for all peracetic acid bleached samples, and this relationship exists on pale and medium-dyed samples.

We can conclude that the differences in whiteness, which were visible on samples after scouring and bleaching, remained to a certain extent after dyeing. The lighter fabrics achieved lower K/S values relative to darker fabrics when dyed under same conditions. E* values reveal that bleaching with hydrogen peroxide enabled the achievement of equal pale and medium colours on alkaline and bioscoured samples, while the colour differences between alkaline and bioscoured samples were higher when the samples were bleached with peracetic acid and dyed. The colour of one-bath pretreated and dyed fabrics was close to the colour of alkaline scoured, peracetic acid bleached and dyed sample, which confirms that hydrogen peroxide bleaching covered the differences in colour arising from different scouring methods, while peracetic acid bleaching preserved those differences.

The colour evenness was excellent for all samples. The standard deviation of ten colour difference values D*E** was below 0.06 for all dyed samples. We can conclude that all presented types of pretreatment are appropriate for further dyeing with reactive dyes, but the initial colour of the material should be considered when the dyeing recipes are prepared.

ECOLOGICAL PARAMETERS

Conventional treatment of cotton fibres was conducted in an alkaline environment: final pH at alkaline scouring and at bleaching with hydrogen peroxide was around 12.5. Such alkaline baths should be neutralised prior to drainage into the sewage system. At neutralisation, salts that additionally load wastewaters are produced. While bleaching with peracetic acid and at both combined processes, the final pH value of the bath was near 6. Since neither of these processes requires neutralisation of fibres, the treatment process can be shorter and less expensive.

While scouring with pectinases and bleaching with peracetic acid, the consumption of energy required to heat the bath was also lower. Conventional processes of scouring and bleaching were performed at temperatures near the boiling point, whereas bioscouring and bleaching with peracetic acid were conducted at a temperature of 55°C. Due to the lower temperature, less energy was required. The consumption of water and energy is the lowest at combined scouring/bleaching treatments. Consequently, at these processes arises the lowest amount of effluents and the produced wastewater is biodegradable.

Enzymatically and alkaline scoured cotton fabrics have similar water absorbency, tenacity at maximum load and degree of polymerisation. Because of lower loss of mass and lower whiteness of enzymatically scoured fabrics noticeable differences in colour occur between differently scoured samples dyed to light hues. The colour differences are overcame at medium and pale shades. During bleaching with hydrogen peroxide the differences in whiteness arising from previous scouring processes disappeared.

All samples obtained high whiteness values. At bleaching with peracetic acid the obtained whiteness values are lower and the differences from previous scouring processes remained visible. This causes that the colour differences between differently scoured samples, which were bleached with hydrogen peroxide prior to dyeing, are not visually perceivable, while they remained visible on samples bleached with peracetic acid at light and medium shade dyeing. At one-bath processes of scouring/bleaching with pectinases and peracetic acid the degree of whiteness of fabrics is higher than at two-step scouring and bleaching with peracetic acid, but lower than at bleaching with hydrogen peroxide. The colour of dyed fabrics is equal to alkaline scoured and peracetic acid bleached fabric. The different shades obtained on differently pretreated fabrics are the consequence of their different initial whiteness values.

All other parameters are very similar, the exhaustion rate, the fixation rate and the fastness properties. The bioscouring and bleaching with peracetic acid, especially the one-bath processes, are suitable treatments before dyeing with reactive dyes. Lower amount of water and energy is consumed than at conventionally pretreatment and the fibres are not damaged at all. For reproducible dyeing the initial colour of the fabric should be considered when preparing the dyeing recipes.

7

Economic Benefit of Cotton Fibre

Cotton is a soft, fluffy staple Fibre that grows in a boll, or protective capsule, around the seeds of cotton plants of the genus Gossypium in the family of Malvaceae. The Fibre is almost pure cellulose. Under natural conditions, the cotton bolls will tend to increase the dispersion of the seeds.

The plant is a shrub native to tropical and subtropical regions around the world, including the Americas, Africa, and India. The greatest diversity of wild cotton species is found in Mexico, followed by Australia and Africa. Cotton was independently domesticated in the Old and New Worlds. The English name derives from the Arabic (al) qumn which began to be used circa 1400 AD.

The Fibre is most often spun into yarn or thread and used to make a soft, breathable textile. The use of cotton for fabric is known to date to prehistoric times; fragments of cotton fabric dated from 5000 BC have been excavated in Mexico and the Indus Valley Civilization (modern-day Pakistan and some parts of India). Although cultivated since antiquity, it was the invention of the cotton gin that lowered the cost of production that led to its widespread use, and it is the most widely used natural Fibre cloth in clothing today.

Current estimates for world production are about 25 million tonnes or 110 million bales annually, accounting for 2.5per cent of the world's arable land. China is the world's largest producer of cotton, but most of this is used domestically. The United States has been the largest exporter for many years. In the United States, cotton is usually measured in bales, which measure approximately 0.48 cubic metres (17 cubic feet) and weigh 226.8 kilograms (500 pounds).

TYPES

There are four commercially grown species of cotton, all domesticated in antiquity:

- Gossypium hirsutum – upland cotton, native to Central America, Mexico, the Caribbean and southern Florida, (90per cent of world production)
- Gossypium barbadense – known as extra-long staple cotton, native to tropical South America (8per cent of world production)

- Gossypium arboreum – tree cotton, native to India and Pakistan (less than 2per cent)
- Gossypium herbaceum – Levant cotton, native to southern Africa and the Arabian Peninsula (less than 2per cent)

The two New World cotton species account for the vast majority of modern cotton production, but the two Old World species were widely used before the 1900s. While cotton fibers occur naturally in colors of white, brown, pink and green, fears of contaminating the genetics of white cotton have led many cotton-growing locations to ban the growing of colored cotton varieties, which remain a specialty product.

GOSSYPIUM HIRSUTUM

Gossypium hirsutum, also known as upland cotton or Mexican cotton, is the most widely planted species of cotton in the United States, constituting some 95per cent of all cotton production there. It is native to Mexico, the West Indies, northern South America, Central America and possibly tropical Florida. Worldwide, the figure is about 90per cent of all cotton production is of cultivars derived from this species.

Archeological evidence from the Tehuacan Valley in Mexico shows the cultivation of this species as long ago as 3,500 BC, although there is as yet no evidence as to exactly where it may have been first domesticated. This is the earliest evidence of cotton cultivation in the Americas found thus far.

Gossypium hirsutum includes a number of varieties or cross-bred cultivars with varying Fibre lengths and tolerances to a number of growing conditions. The longer length varieties are called "long staple upland" and the shorter length varieties are referred to as "short staple upland". The long staple varieties are the most widely cultivated in commercial production. Besides being fibre crops, Gossypium hirsutum and Gossypium herbaceum are the main species used to produce cottonseed oil. The Zuni people use this plant to make ceremonial garments, and the fuzz is made into cords and used ceremonially.

GOSSYPIUM BARBADENSE

Gossypium barbadense, also known as extra long staple (ELS) cotton as it generally has a staple of at least 1 3/8" or longer, is a species of cotton plant. Varieties of ELS cotton include American Pima, Egyptian Giza, Indian Suvin, Chinese Xinjiang, Sudanese Barakat, and Russian Tonkovoloknistyi. It is a tropical, frost-sensitive perennial plant that produces yellow flowers and has black seeds. It grows as a small, bushy tree and yields cotton with unusually long, silky fibers. To grow, it requires full sun and high humidity and rainfall.

This plant contains the chemical gossypol, which reduces its susceptibility to insect and fungal damage. In Suriname's traditional medicine, the leaves of G. barbadense are used to treat hypertension and delayed/irregular menstruation.

The name Pima was applied in Honour of the Pima Indians, who helped raise the cotton on USDA experimental farms in Arizona in the early 1900s. The first clear sign of domestication of this cotton species comes the Early Valdivia phase site of Real Alto on the coast of Ecuador (4400 BC) and from Ancon, a site on the Peruvian coast, where cotton bolls dating to 4200 BC were found. By 1000 BC, Peruvian cotton bolls were indistinguishable from modern cultivars of G. barbadense. Cotton growing became widespread in South America and spread to the West Indies, where Christopher Columbus encountered it. Cotton became a commercial plantation crop tended by slaves in the West Indies, so that by the 1650s, Barbados had become the first British West Indies colony to export cotton.In about 1786, planting of Sea Island cotton began in the former British North American colonies, on the Sea Islands of South Carolina and Georgia, when cotton planters were brought over from Barbados. Among the earliest planters of Sea Island cotton in America was an Englishman, Francis Levett, who later fled his Georgia plantation at the outbreak of the American Revolution and went to the Bahamas, where he attempted to introduce cotton production, but failed. Sea Island cotton commanded the highest price of all the cottons, due to its long staple (1.5 to 2.5 inches, 35 to 60 mm) and its silky texture; it was used for the finest cotton counts and often mixed with silk. It was also grown on the uplands of Georgia, where the quality was inferior, and was soon surpassed in commercial production by another native American species, upland cotton (Gossypium hirsutum), which today represents about 95per cent of U.S. production.The term Egyptian cotton is usually applied to the extra long staple cotton produced in Egypt and used by luxury and upmarket brands worldwide. It also has the most upper thread count.

American Pima accounts for less than 5per cent of U.S. cotton production. It is grown chiefly in California, with small acreages in West Texas, New Mexico and Arizona.

For purposes of federal support, the 2002 farm bill (P.L. 101-171, Sec. 1001) defined ELS cotton.

ELS cotton, like upland cotton, is eligible for marketing assistance loans and loan deficiency payments (LDPs). The national loan rate for ELS cotton under the 2002 farm bill was $0.7977 per pound. ELS cotton, in contrast to upland cotton, does not qualify for direct payments or counter-cyclical payments.

GOSSYPIUM ARBOREUM

Gossypium arboreum, commonly called tree cotton, is a species of cotton native to India, Pakistan and other tropical and subtropical regions of the Old World. There is evidence of its cultivation as long ago as the Harappan civilization of the Indus Valley for the production of cotton textiles. This species of cotton was also introduced into East Africa and was grown by the Meroe civilization in Nubia. The shrub was included in Linnaeus's Species Plantarum published in 1753. The holotype was also supplied by him, which is now in the Linnean Herbarium in the Swedish Museum of Natural History. Tree cotton is a shrub attaining heights of one to two metres. Its branches are covered with pubescence and are purple in colour. Stipules are present at the leaf base and they are linear to lanceolate in shape and sometimes falcate (*i.e.* sickle-shaped). The leaves are attached to the stem by a 1.5 to 10 cm petiole. The blades are ovate to orbicular in shape and have five to seven lobes, making them superficially resemble a maple leaf. The lobes are linear to lanceolate, and often a tooth is present in the sinus. Glands are present along the midrib or occasionally on the adjacent nerves. The leaves are glabrescent, meaning the pubescence is lost with age, but when it is present on young leaves, it is both stellate (*i.e.* star-shaped) and simple.

The flowers are set on short pedicels (*i.e.* flower stalks). An epicalyx is present, which is a series of subtending bracts that resemble sepals. Its large, ovate segments are dentate (*i.e.* toothed along the margins), though sometimes only very slightly so. They are cordate (*i.e.* heart-shaped) at the base and acute at the apex. The true calyx is small, measuring only about 5 mm in length. Its shape is cupular, and five subtle dentations are present. The corolla is a pale yellow on colour, sometimes with a purple centre, and occasionally entirely purple. It measures 3 to 4 cm in length. The staminal tube bears the anthers and is 1.5 to 2 cm in length. The fruit is a three- or four-celled capsule measuring 1.5 to 2.5 cm across. It is ovoid or oblong in shape and glabrous (*i.e.* hairless). The surface is pitted and a beak is present at the terminal end. The seeds within are globular and are covered in long white cotton.

GOSSYPIUM HERBACEUM

Gossypium herbaceum, commonly known as Levant cotton, is a species of cotton native to the semi-arid regions of sub-Saharan Africa and Arabia where it still grows in the wild as a perennial shrub. It is a sister-species of Gossypium arboreum.

It was first cultivated in Western Sudan, from there it spread to India, before being introduced to Egypt. It reached China in the 700 AD and was first cultivated from this period. It can also be found in Pakistan and almost all of the former Soviet Union areas.

A legend was perpetuated from a factual description of this plant by Greek historian Herodotus in the 5th century BC. Although his book, simply titled Histories, was an account of a war between the Persian Empire and the Greek city-states. It also contained descriptions of vast lands beyond the boundaries of the world known by the Greeks at that time. He wrote: "certain trees...bear forth their fruit fleeces surpassing those of sheep in beauty and excellence, and the natives clothe themselves in cloths made therefrom." From this description came the legend of the "vegetable lamb plant" which was said to be a real sheep. The tree would grow from a melon-like seed and grow into a lamb rooted to the earth by a stem from its navel. It was said to graze on the surrounding vegetation until the all greenery around it was devoured at which point it would wither and die. A 14th-century traveler by the name of Sir John Mandeville, professed to eating the flesh of this herbal beast. Although scientist tried to debunk this tale it was not officially labeled as a fable until 1887.

G. herbaceum has high stems that grow 2 feet (0.61 m) to 6 feet (1.8 m) high with wide, hairy leaves. Their flowers are small and yellow with a purple center. When ripe and in warm weather, the flower capsule will burst and expose the cotton surrounding the seeds firmly. The cotton produced by this plant is short, about 2 inches (5.1 cm) long and is firmly attached to the seed, which is covered in hairy down. An acre of cotton can be expected to produce about 300 pounds (140 kg).

Cotton is usually used as a textile while making clothing and can be made into yarns and sheets of fabric. In the Levant seeds are often used for as food. It is utilized so often because of its comfortable, breathable properties. It has been cultivated for women's menstrual cycle pains and irregular bleeding, and it also has been used after birth to expel placenta afterbirth and to increase lactation. Cotton has been used for gastrointestinal issues also, such as hemorrhages, nausea, and diarrhea, as well as fevers and headaches, especially in the southern United States. Levant cotton seed extract, gossypol, also has a potential use as a male contraceptive but can cause infertility after discontinuing. In lab rat studies, it has been able to stop pregnancies early. Orally administered ethyl ether and ethanol extracts of Gossypium herbaceum significantly decreased the blood glucose level. Gossypium herbaceum is not only lowered TC, TG, LDL, VLDL levels but also increased level of cardioprotective lipid HDL Therefore, Gossypium herbaceum has potential role to prevent formation of atherosclerosis and coronary heart disease. The study demonstrated that both above given extracts of Gossypium herbaceum could be useful in management of diabetes associated with abnormalities in lipid profiles.

HISTORY

Cotton was used in the Old World at least 7,000 years ago (5th millennium BC). Evidence of cotton use has been found at the site of Mehrgarh, where early cotton threads have been preserved in copper beads. Cotton cultivation became more widespread during the Indus Valley Civilization, which covered parts of modern eastern Pakistan and northwestern India. The Indus cotton industry was well developed and some methods used in cotton spinning and fabrication continued to be used until the industrialization of India. Between 2000 and 1000 BC cotton became widespread across much of India. For example, it has been found at the site of Hallus in Karnataka dating from around 1000 BC.

Cotton fabrics discovered in a cave near Tehuacán, Mexico have been dated to around 5800 BC, although it is difficult to know for certain due to Fibre decay. Other sources date the domestication of cotton in Mexico to approximately 5000 to 3000 BC.

The Greeks and the Arabs were not familiar with cotton until the Wars of Alexander the Great, as his contemporary Megasthenes told Seleucus I Nicator of "there being trees on which wool grows" in "Indica". This might actually be a reference to the 'tree cotton', Gossypium arboreum, which is a native of the Indian subcontinent.

According to the Columbia Encyclopaedia:

- Cotton has been spun, woven, and dyed since prehistoric times. It clothed the people of ancient India, Egypt, and China. Hundreds of years before the Christian era, cotton textiles were woven in India with matchless skill, and their use spread to the Mediterranean countries.

In Iran (Persia), the history of cotton dates back to the Achaemenid era (5th century BC); however, there are few sources about the planting of cotton in pre-Islamic Iran. The planting of cotton was common in Merv, Ray and Pars of Iran. In the poems of Persian poets, especially Ferdowsi's Shahname, there are references to cotton ("panbe" in Persian). Marco Polo (13th century) refers to the major products of Persia, including cotton. John Chardin, a French traveler of 17th century, who had visited the Safavid Persia, has approved the vast cotton farms of Persia.

During the Han dynasty, cotton was grown by Non- Chinese peoples in the southern Chinese province of Yunnan.

In Peru, cultivation of the indigenous cotton species Gossypium barbadense was the backbone of the development of coastal cultures, such as the Norte Chico, Moche and Nazca. Cotton was grown upriver, made into nets and traded with fishing villages along the coast for large supplies of fish. The Spanish who came to Mexico and Peru in the early 16th century found the people growing cotton and wearing clothing made of it.

During the late medieval period, cotton became known as an imported Fibre in northern Europe, without any knowledge of how it was derived, other than that it was a plant. Because Herodotus had written in his Histories, Book III, 106, that in India trees grew in the wild producing wool, it was assumed that the plant was a tree, rather than a shrub. This aspect is retained in the name for cotton in several Germanic languages, such as German Baumwolle, which translates as "tree wool" (Baum means "tree"; Wolle means "wool"). Noting its similarities to wool, people in the region could only imagine that cotton must be produced by plant-borne sheep. John Mandeville, writing in 1350, stated as fact the now-preposterous belief: "There grew there [India] a wonderful tree which bore tiny lambs on the endes of its branches. These branches were so pliable that they bent down to allow the lambs to feed when they are hungrie. By the end of the 16th century, cotton was cultivated throughout the warmer regions in Asia and the Americas.

India's cotton-processing sector gradually declined during British expansion in India and the establishment of colonial rule during the late 18th and early 19th centuries. This was largely due to aggressive colonialist mercantile policies of the British East India Company, which made cotton processing and manufacturing workshops in India uncompetitive. Indian markets were increasingly forced to supply only raw cotton and were forced, by British-imposed law, to purchase manufactured textiles from Britain.

INDUSTRIAL REVOLUTION IN BRITAIN

The advent of the Industrial Revolution in Britain provided a great boost to cotton manufacture, as textiles emerged as Britain's leading export. In 1738, Lewis Paul and John Wyatt, of Birmingham, England, patented the roller spinning machine, and the flyer-and-bobbin system for drawing cotton to a more even thickness using two sets of rollers that traveled at different speeds. Later, the invention of the James Hargreaves' spinning jenny in 1764, Richard Arkwright's spinning frame in 1769 and Samuel Crompton's spinning mule in 1775 enabled British spinners to produce cotton yarn at much higher rates. From the late 18th century on, the British city of Manchester acquired the nickname "Cottonopolis" due to the cotton industry's omnipresence within the city, and Manchester's role as the heart of the global cotton trade.

Production capacity in Britain and the United States was improved by the invention of the cotton gin by the American Eli Whitney in 1793. Before the development of cotton gins, the cotton fibers had to be pulled from the seeds tediously by hand. By the late 1700s a number of crude ginning machines had been developed. However, to produce a bale of cotton required over 600 hours of human Labour, making large scale production uneconomical in the United States, even with the use of humans as slave Labour. The gin that Whitney manufactured (the Holmes design) reduced the hours down to just a dozen or so per bale. Although Whitney patented his own design for a cotton gin, he manufactured a prior design from Henry Odgen Holmes, for which Holmes filed a patent in 1796. Improving technology and increasing control of world markets allowed British traders to develop a commercial chain in which raw cotton fibers were (at first) purchased from colonial plantations, processed into cotton cloth in the mills of Lancashire, and then exported on British ships to captive colonial markets in West Africa, India, and China (via Shanghai and Hong Kong).

By the 1840s, India was no longer capable of supplying the vast quantities of cotton fibers needed by mechanized British factories, while shipping bulky, low-price cotton from India to Britain was time-consuming and expensive. This, coupled with the emergence of American cotton as a superior type (due to the longer, stronger fibers of the two domesticated native American species, Gossypium hirsutum and Gossypium barbadense), encouraged British traders to purchase cotton from plantations in the United States and plantations in the Caribbean. By the mid-19th century, "King Cotton" had become the backbone of the southern American economy. In the United States, cultivating and harvesting cotton became the leading occupation of slaves.

During the American Civil War, American cotton exports slumped due to a Union blockade on Southern ports, and also because of a strategic decision by the Confederate government to cut exports, hoping to force Britain to recognize the Confederacy or enter the war. This prompted the main purchasers of cotton, Britain and France, to turn to Egyptian cotton. British and French traders invested heavily in cotton plantations.

The Egyptian government of Viceroy Isma'il took out substantial loans from European bankers and stock exchanges. After the American Civil War ended in 1865, British and French traders abandoned Egyptian cotton and returned to cheap American exports, sending Egypt into a deficit spiral that led to the country declaring bankruptcy in 1876, a key factor behind Egypt's occupation by the British Empire in 1882.

During this time, cotton cultivation in the British Empire, especially India, greatly increased to replace the lost production of the American South. Through tariffs and other restrictions, the British government discouraged the production of cotton cloth in India; rather, the raw Fibre was sent to England for processing. The Indian Mahatma Gandhi described the process:

1. English people buy Indian cotton in the field, picked by Indian Labour at seven cents a day, through an optional monopoly.
2. This cotton is shipped on British ships, a three-week journey across the Indian Ocean, down the Red Sea, across the Mediterranean, through Gibraltar, across the Bay of Biscay and the Atlantic Ocean to London. One hundred per cent profit on this freight is regarded as small.
3. The cotton is turned into cloth in Lancashire. You pay shilling wages instead of Indian pennies to your workers. The English worker not only has the advantage of better wages, but the steel companies of England get the profit of building the factories and machines. Wages; profits; all these are spent in England.
4. The finished product is sent back to India at European shipping rates, once again on British ships. The captains, officers, sailors of these ships, whose wages must be paid, are English. The only Indians who profit are a few lascars who do the dirty work on the boats for a few cents a day.

5. The cloth is finally sold back to the kings and landlords of India who got the money to buy this expensive cloth out of the poor peasants of India who worked at seven cents a day. (Fisher 1932 pp 154–156)

In the United States, Southern cotton provided capital for the continuing development of the North. The cotton produced by enslaved African Americans not only helped the South, but also enriched Northern merchants. Much of the Southern cotton was trans-shipped through northern ports.

Cotton remained a key crop in the Southern economy after emancipation and the end of the Civil War in 1865. Across the South, sharecropping evolved, in which free black farmers and landless white farmers worked on white-owned cotton plantations of the wealthy in return for a share of the profits. Cotton plantations required vast Labour forces to hand-pick cotton.

It was not until the 1950s that reliable harvesting machinery was introduced (prior to this, cotton-harvesting machinery had been too clumsy to pick cotton without shredding the fibers). During the first half of the 20th century, employment in the cotton industry fell, as machines began to replace laborers and the South's rural Labour force dwindled during the World Wars.

Cotton remains a major export of the southern United States, and a majority of the world's annual cotton crop is of the long-staple American variety.

DISORDERS OF COTTON

Potash hunger of cotton was known as rust or black rust before 1892, when George F. Atkinson demonstrated that it could be controlled by additions of potash to the soil. He called the disease yellow leaf blight, but the name rust has remained in common usage. Scientists frequently use the term "rust or potash hunger" because it describes both the symptoms and the cause.

Rust occurs commonly on the lighter soils of the Coastal Plain areas of the Cotton Belt and in many other soils of several mid-South States, including Alabama, Arkansas, Louisiana, and Mississippi. W. W. Gilbert estimated the average damage for the entire Cotton Belt at 4 or 5 percent in 1920, but greater use of balanced fertilisers in the affected areas undoubtedly has reduced the loss in recent years.

Affected plants are usually stunted and fail to develop a normal green colour. The lower or older leaves are first to show the typical mottled appearance; the areas between the veins turn yellow. Towards midseason the symptoms are visible over the entire plant.

The yellowish spots on the leaves enlarge and become reddish brown or bronzed. Later the entire leaf may become blackened, curled downward, and ragged. Brown or black circular or irregular spots are often produced on many affected leaves by the growth of Alternaria and perhaps other secondary invaders. Leaves usually are shed prematurely and often the stalk is left bare. The number and the sise of bolls are reduced. Yields are severely lowered.

Affected bolls do not flare open properly, and so the cotton is hard to pick. Lint and seed from affected plants are inferior. The primary cause of rust is insufficient potash to meet the nutritional requirements of the plant. The condition usually is found on potash-deficient soils, but not necessarily so. Excessive applications of nitrogen or phosphorus, or both, often may accentuate severity.

Over-liming or a scarcity of sodium, which substitutes to some extent for potash, likewise tend to increase rust. Lack of humus in the soil and improper drainage have also been listed as conditions responsible for or contributing to severity of the disease.

The application of fertilisers that contain ample potash to furnish the needs of the plant throughout the season is considered the standard method of control. The exact amount needed to correct the condition depends on the supply of other soil nutrients.

A common practice on lighter soils, where early rains can easily leach out some of the available potash supplied before planting, is to apply some potash along with nitrogen as a side dressing after the plants are established.

Potash hunger should not be confused with the true rust of cotton, which is caused by attack of the parasitic fungus Puccinia stakmanii, and occurs in limited areas of the Western States. Fusarium wilt and root knot nematodes also are often prevalent in the soils where potash hunger is common. It is not always easy to determine the exact degree of damage associated with each cause in this complex.

The Crazy Top disorder of cotton was first recognised by C. J. King and H. F. Loomis in a field of Pima cotton near Scottsdale, Ariz., in 1919. Three years later it was observed in a field of upland cotton near Casa Grande, Ariz. By 1924 the disorder had become common in Arizona and by 1936 it was still considered one of the most important problems of cotton production in certain districts of the State. Mild stages of the disease were also known to have appeared in the San Joaquin and Imperial Valleys of California between 1926 and 1933. O. F. Cook was first to present an accurate description of the disorder. He named the disease acromania, but the term crazy top, which is descriptive of the abnormal branching and fruiting in the upper part of affected plants, has remained the one in common use.

The first symptom is an abrupt change in the type of new growth made in the top of the cotton plants. Fruiting branches frequently are replaced by vegetative branches, which have regrown in an upright position and are developing an abnormal or crazy appearance in the tops of the plants. Typical features of the abnormality include reductions and distortions of leaves, internodes, bracts, and floral parts. The leaves are small, rounded, cupped, and thickened. Flowers are small and distorted. Sterility is common and may be expressed either as complete suppression of floral buds or as profuse shedding of flowers. The few bolls that develop are usually small and malformed and contain a reduced number of seed.

The disease occurs only on calcareous soils and has been found almost entirely related to irrigation practice. It is associated with the checking of growth from water shortages and the resumption of growth when abundant moisture is restored. The disease is controlled easily by applying irrigation water frequently enough to prohibit a checking of plant growth during the summer months. Water shortage may not be the only factor involved in bringing about the abnormalities maintaining organic matter at higher levels by rotation with alfalfa is beneficial.

Crinkle Leaf is the name given by D. C. Neal in 1937 to a peculiar disorder of cotton plants occurring in Lintonia and Oliver silt loam soils of Louisiana and Arkansas. The typical symptoms are a puckering, mottling, partial chlorosis, and distortion of the leaves. The branches are often fasciated and the floral buds, flowers, and bolls are distorted. The fibre from affected bolls is often so weak as to be considered worthless.

Dr. Neal and H. C. Lovett demonstrated by experiments they conducted in Louisiana in 1937 and 1938 that the disease was associated with high soil acidity, calcium deficiency, and manganese toxicity. The typical symptoms were readily produced by adding increasing amounts of manganese sulfate to pot cultures of cotton. But it was easy to control the disease in the field by adding limestone. At the higher pH produced by treating the soil with limestone, or with other material containing basic carbonates, the manganese apparently is precipitated, and so the plants remain healthy.

THRIPS INJURY is not strictly a nonparasitic disease of cotton, but I include a brief description because its symptoms are not unlike those of certain stages of crinkle leaf and crazy top.

Infestation by thrips frequently causes serious injury to cotton plants. Reduced stands or reduced yields result from stunting and delayed fruiting of the plants. Damage is severe only in the early-season development of the plants and is usually restricted to mutilation of seedling leaves as they unfold from the bud. Often, however, the terminal bud is destroyed, so that the plant gets a stunted appearance. Until it outgrows the disorder around mid-season, the branching is bunched or fasciated. The leaves may appear torn, perforated, cupped, and distorted. Usually there will be little if any change from normal green colour.

A serious aspect of the disease is that the injury and decreased vigour tend to make plants more susceptible to the sore shin disease of cotton seedlings caused by Rhizoctoma solani a fact demonstrated in experiments conducted by Neal and L. D. Newsome in Louisiana in 1950.

Considerable variation in susceptibility among commercial varieties of cotton was observed by W. W. Ballard in Georgia. He pointed out that dense pubescence of juvenile terminal leaves appeared to be associated with resistance to thrips in some cases. He noted also that the degree of injury was usually correlated with the amount of aborted terminal buds, although a few varieties exhibited a degree of injury independent of the proportion of terminal buds aborted.

2,4-D (2,4-dichlorophenoxyacetic acid), a hormone-type herbicide commonly used to kill broadleaf pests in lawns, is extremely toxic to cotton plants.

The minute quantities that light on cotton plants when the drift from adjacent applications is not controlled can cause severe injury. The danger has been thoroughly recognised since this weed killer was first introduced commercially in 1946 but, despite adequate and repeated warnings, cotton may be found injured occasionally from drift, from accidental contamination of other chemicals used on cotton, or from improperly cleaned machinery and containers that have been used with 2,4-D.

The symptoms on cotton are characteristic and specific formative effects, except where the dosage is sufficiently high to kill the plants or plant parts outright. The effects are first noted near the growing points and the characteristic symptoms are usually the result of an abnormal stimulation of growth of the affected parts. Leaves are greatly modified. They become narrow, closely and conspicuously veined, and deeply lobed. Affected flowers are modified in a manner similar to the leaf elongated and narrow. Bracts are modified. They become deeply lobed and elongated and they fail to separate; they grow as a sheath around the developing boll.

When enough material is absorbed to kill or inhibit the terminal bud, the plant may become branched and fasciated, with slender and malformed leaves, branches, buds, and flowers. Affected bolls and squares may turn yellow and then drop, although some malformed and partially fertile mature bolls have been observed. Often, however, the squares or bolls die without shedding.

Excessive shedding of leaves as a result of 2,4-D injury has not been reported. If the tops of plants are damaged, they may send out strong laterals, which produce apparently normal branches. The fully grown leaves, branches, or bolls seldom show much injury from the low dosages.

The formative effects in cotton continue to develop for some weeks following application of 2,4-D but the persistence of the stimulus depends on the stage of plant development and the concentration applied. If young bolls are present, the seed may be, injured. The likelihood of damage is greater for seed developing from plants treated at the time of flowering and, normally, no effects on the seed will be noted if parent plants are treated before the young square stage of growth.

Some instances have been reported where applications of organic insecticides have been associated with injury to cotton plants similar to that produced by 2,4-D. Many are undoubtedly cases of contamination, but with the use of certain of the organic phosphates it may be the insecticide itself causing the injury. Wayne J. McIlrath has shown a striking similarity of symptoms produced by treating comparable plants with 2,4-D and with commercial HETP (hexaethyl tetra-phosphate), and to some extent with TEPP (tetraethyl pyrophosphate).

POLYMER FIBERS

Polymer fibers are a subset of man-made fibers, which are based on synthetic chemicals (often from petrochemical sources) rather than arising from natural materials by a purely physical process. These fibers are made from:

- polyamide nylon
- PET or PBT polyester
- phenol-formaldehyde (PF)
- polyvinyl chloride Fibre (PVC) vinyon
- polyolefins (PP and PE) olefin Fibre
- acrylic polyesters, pure polyester PAN fibers are used to make carbon Fibre by roasting them in a low oxygen environment. Traditional acrylic Fibre is used more often as a synthetic replacement for wool. Carbon fibers and PF fibers are noted as two resin-based fibers that are not thermoplastic, most others can be melted.
- aromatic polyamids (aramids) such as Twaron, Kevlar and Nomex thermally degrade at high temperatures and do not melt. These fibers have strong bonding between polymer chains
- polyethylene (PE), eventually with extremely long chains/ HMPE (*e.g.* Dyneema or Spectra).
- Elastomers can even be used, *e.g.* spandex although urethane fibers are starting to replace spandex technology.
- polyurethane Fibre
- Elastolefin

Coextruded fibers have two distinct polymers forming the Fibre, usually as a core-sheath or side-by-side. Coated fibers exist such as nickel-coated to provide static elimination, silver-coated to provide anti-bacterial properties and aluminum-coated to provide RF deflection for radar chaff. Radar chaff is actually a spool of continuous glass tow that has been aluminum coated. An aircraft-mounted high speed cutter chops it up as it spews from a moving aircraft to confuse radar signals.

ACRYLIC FIBRE

Acrylic fibers are synthetic fibers made from a polymer (polyacrylonitrile) with an average molecular weight of ~100,000, about 1900 monomer units. To be called acrylic in the U.S, the polymer must contain at least 85per cent acrylonitrile monomer. Typical comonomers are vinyl acetate or methyl acrylate. DuPont created the first acrylic fibers in 1941 and trademarked them under the name Orlon. Acrylic is also called acrilan fabric. It was first developed in the mid-1940s but was not produced in large quantities until the 1950s. Strong and warm, acrylic Fibre is often used for sweaters and tracksuits and as linings for boots and gloves, as well as in furnishing fabrics and carpets. It is manufactured as a filament, then cut into short staple lengths similar to wool hairs, and spun into yarn.

Modacrylic is a modified acrylic Fibre that contains at least 35per cent and at most 85per cent acrylonitrile monomer. The comonomers vinyl chloride, vinylidene chloride or vinyl bromide used in modacrylic give the Fibre flame retardant properties. End-uses of modacrylic include faux fur, wigs, hair extensions and protective clothing.

Production

The polymer is formed by free-radical polymerization in aqueous suspension. The Fibre is produced by dissolving the polymer in a solvent such as N,N-dimethylformamide or aqueous sodium thiocyanate, metering it through a multi-hole spinnerette and coagulating the resultant filaments in an aqueous solution of the same solvent (wet spinning) or evaporating the solvent in a stream of heated inert gas (dry spinning). Washing, stretching, drying and crimping complete the processing. Acrylic fibers are produced in a range of deniers, typically from 0.9 to 15, as cut staple or as a 500,000 to 1 million filament tow. End uses include sweaters, hats, hand-knitting yarns, socks, rugs, awnings, boat covers, and upholstery; the Fibre is also used as "PAN" precursor for carbon Fibre. Production of acrylic fibers is centered in the Far East, Turkey, India, Mexico, and South America, though a number of European producers still continue to operate, including Dralon and Fisipe. US producers have ended production, though acrylic tow and staple are still spun into yarns in the USA. Former U.S. brands of acrylic were Acrilan (Monsanto), Creslan (American Cyanamid), and Orlon (DuPont). Other brand names that are still in use include Dralon (Dralon GmbH).

Textile uses

Acrylic is lightweight, soft, and warm, with a wool-like feel. It can also be made to mimic other fibers, such as cotton, when spun on short staple equipment. Some acrylic is extruded in colored or pigmented form; other is extruded in "ecru", otherwise known as "natural," "raw white," or "undyed." Pigmented Fibre has highest light-fastness. Its fibers are very resilient compared to both other synthetics and natural fibers. Some acrylic is used in clothing as a less expensive alternative to cashmere, due to the similar feeling of the materials. Some acrylic fabrics may fuzz or pill easily. Other fibers and fabrics are designed to minimize pilling. Acrylic takes Colour well, is washable, and is generally hypoallergenic. End-uses include socks, hats, gloves, scarves, sweaters, home furnishing fabrics, and awnings.

Acrylic is resistant to moths, oils, chemicals, and is very resistant to deterioration from sunlight exposure.

Acrylic is the "workhorse" hand-crafting Fibre for crafters who knit or crochet; acrylic yarn may be perceived as "cheap" because it is typically priced lower than its natural-Fibre counterparts, and because it lacks some of their properties, including softness and the ability to felt or take acid dyes. The Fibre requires heat to "kill" or set the shape of the finished garment, and it isn't as warm as alternatives like wool. Some knitters also complain that the Fibre "squeaks" when knitted, or that it is painful to knit with because of a lack of "give" or stretch in the yarn. On the other hand, it can be useful in certain items, like garments for babies, which require constant washing, because it is machine-washable and extremely Colour-fast. Acrylic can irritate the skin of people with dermatological conditions such as eczema.

ARAMID

Aramid fibers are a class of heat-resistant and strong synthetic fibers. They are used in aerospace and military applications, for ballistic rated body armor fabric and ballistic composites, in bicycle tires, and as an asbestos substitute. The name is a portmanteau of "aromatic polyamide". They are fibers in which the chain molecules are highly oriented along the Fibre axis, so the strength of the chemical bond can be exploited.

Aromatic polyamides were first introduced in commercial applications in the early 1960s, with a meta-aramid Fibre produced by DuPont as HT-1 and then under the trade name Nomex. This Fibre, which handles similarly to normal textile apparel fibers, is characterized by its excellent resistance to heat, as it neither melts nor ignites in normal levels of oxygen. It is used extensively in the production of protective apparel, air filtration, thermal and electrical insulation as well as a substitute for asbestos. Meta-aramid is also produced in the Netherlands and Japan by Teijin under the trade name Conex, In Korea by Toray under the trade name Arawin, in China by Yantai Tayho under the trade name New Star, by SRO Group (China) under the trade name X-Fiper, and a variant of meta-aramid in France by Kermel under the trade name Kermel.

Based on earlier research by Monsanto Company and Bayer, a Fibre – para-aramid – with much higher tenacity and elastic modulus was also developed in the 1960s–1970s by DuPont and Akzo Nobel, both profiting from their knowledge of rayon, polyester and nylon processing. Much work was done by Stephanie Kwolek in 1961 while working at DuPont, and that company was the first to introduce a *para-aramid* called Kevlar in 1973. A similar Fibre called Twaron with roughly the same chemical structure was introduced by Akzo in 1978. Due to earlier patents on the production process, Akzo and DuPont engaged in a patent dispute in the 1980s. Twaron is currently owned by the Teijin company.

Para-aramids are used in many high-tech applications, such as aerospace and military applications, for "bullet-proof" body armor fabric.

The Federal Trade Commission definition for aramid Fibre is:

- A manufactured Fibre in which the Fibre-forming substance is a long-chain synthetic polyamide in which at least 85per cent of the amide linkages, (-CO-NH-) are attached directly to two aromatic rings.

Health

During the 1990s, an *in vitro* test of aramid fibers showed they exhibited "many of the same effects on epithelial cells as did asbestos, including increased radiolabeled nucleotide incorporation into DNA and induction of ODC (ornithine decarboxylase) enzyme activity", raising the possibility of carcinogenic implications. However, in 2009, it was shown that inhaled aramid fibrils are shortened and quickly cleared from the body and pose little risk.

Production

World capacity of para-aramid production is estimated at about 41,000 tonnes/year in 2002 and increases each year by 5–10per cent. In 2007 this means a total production capacity of around 55,000 tonnes/year.

Polymer preparation

Aramids are generally prepared by the reaction between an amine group and a carboxylic acid halide group. Simple AB homopolymers may look like:

$n\text{NH}_2\text{-Ar-COCl}$! $\text{-(NH-Ar-CO)}_n\text{-} + n\text{HCl}$

The most well-known aramids (Kevlar, Twaron, Nomex, New Star and Teijinconex) are AABB polymers. Nomex, Teijinconex and New Star contain predominantly the meta-linkage and are poly-*metaphenylene isophthalamide*s (MPIA). Kevlar and Twaron are both *p*-phenylene terephthalamides (PPTA), the simplest form of the AABB para-polyaramide. PPTA is a product of *p*-phenylene diamine (PPD) and terephthaloyl dichloride (TDC or TCl). Production of PPTA relies on a co-solvent with an ionic component (calcium chloride ($CaCl_2$)) to occupy the hydrogen bonds of the amide groups, and an organic component (N-methyl pyrrolidone (NMP)) to dissolve the aromatic polymer. Prior to the invention of this process by Leo Vollbracht, who worked at the Dutch chemical firm Akzo, no practical means of dissolving the polymer was known. The use of this system led to an extended patent dispute between Akzo and DuPont.

Spinning lh3)

After production of the polymer, the aramid Fibre is produced by spinning the dissolved polymer to a solid Fibre from a liquid chemical blend. Polymer solvent for spinning PPTA is generally 100per cent anhydrous sulfuric acid (H_2SO_4).

Appearances

- Fibre
- Chopped Fibre
- Powder
- Pulp

Other types of aramids

Besides meta-aramids like Nomex, other variations belong to the aramid Fibre range. These are mainly of the copolyamide type, best known under the brand name Technora, as developed by Teijin and introduced in 1976. The manufacturing process of Technora reacts PPD and 3,4'-diaminodiphenylether (3,4'-ODA) with terephthaloyl chloride (TCl). This relatively simple process uses only one amide solvent and therefore spinning can be done directly after the polymer production.

Aramid Fibre characteristics

Aramids share a high degree of orientation with other fibers such as ultra high molecular weight polyethylene, a characteristic that dominates their properties.

General

- good resistance to abrasion
- good resistance to organic solvents
- Non-conductive
- no melting point, degradation starts from 500°C
- low flammability
- good fabric integrity at elevated temperatures
- sensitive to acids and salts
- sensitive to ultraviolet radiation
- prone to static build-up unless finished

Para-aramids

- para-aramid fibers such as Kevlar and Twaron, provide outstanding strength-to-weight properties
- high Young's modulus
- high tenacity
- low creep
- low elongation at break (~3.5per cent)
- difficult to dye – usually solution dyed

Major industrial uses

- flame-resistant clothing (example military MIL-G-181188B suits).
- heat protective clothing and helmets
- body armor, competing with PE based Fibre products such as Dyneema and Spectra
- composite materials
- asbestos replacement (*e.g.* brake linings)
- hot air filtration fabrics
- tires, newly as Sulfron (sulfur modified Twaron)
- mechanical rubber goods reinforcement
- ropes and cables
- wicks for fire dancing
- optical Fibre cable systems
- sail cloth (not necessarily racing boat sails)
- sporting goods
- drumheads
- wind instrument reeds, such as the Fibracell brand
- loudspeaker diaphragms
- boathull material
- Fibre reinforced concrete
- reinforced thermoplastic pipes
- tennis strings (*e.g.* by Ashaway and Prince tennis companies)
- hockey sticks (normally in composition with such materials as wood and carbon)
- snowboards
- jet engine enclosures

TWARON

Twaron is the brandname of Teijin Aramid for a para-aramid. It is a heat-resistant and strong synthetic fibre developed in the early 1970s by the Dutch company AKZO, division ENKA, later Akzo Industrial Fibers. The research name of the para-aramid fibre was originally Fibre X, but it was soon called Arenka. Although the Dutch para-aramid Fibre was developed only a little later than DuPont's Kevlar, introduction of Twaron as a commercial product came much later than Kevlar due to financial problems at the AKZO company in the 1970s.

This is a chronology of the development of Twaron:

- In 1960s a research Programme starts for "Fibre X."
- In 1972 the ENKA Research laboratory develops a para-aramid called Arenka.
- In 1973 Akzo decides to use sulfuric acid (H_2SO_4) as a solvent for spinning.

- In 1976 a pilot plant is built.
- In 1977 first production starts.
- In 1984 the product is renamed Twaron.
- In 1986 commercial production is started at five locations and nine plants.
- In 1987 Twaron is introduced as a commercial product.
- In 1989 the aramid business of Akzo becomes an independent Business Unit called Twaron BV.
- Since 2000 Twaron BV is owned by the Teijin Group, now called Teijin Twaron BV and based in Arnhem, The Netherlands. The main production facilities for Twaron are in Emmen and Delfzijl.
- In 2007 Teijin Twaron expands for the fourth time in six years and also changes its name into Teijin Aramid.

Production

Polymer preparation

Twaron is a *p*-phenylene terephthalamide (P*p*PTA), the simplest form of the AABB para polyaramide. P*p*PTA is a product of *p*-phenylene diamine (PPD) and terephthaloyl dichloride (TDC). To dissolve the aromatic polymer Twaron used a co-solvent of N-methyl pyrrolidone (NMP) and an ionic component (calcium chloride $CaCl_2$) to occupy the hydrogen bonds of the amide groups. Prior to the invention of this process by Leo Vollbracht, working at the Dutch chemical firm AKZO, no practical means of dissolving the polymer was known. The use of this system by DuPont led to a patent war between AKZO and DuPont as Dupont initially used the carcinogenic HMPT (hexamethylphosphoramide). Despite heavy research DuPont now also uses the AKZO patent to use the less hazardous NMP in the Kevlar process.

Spinning

After the production of the Twaron polymer in Delfzijl, the polymer is brought to Emmen, where fibres are produced by spinning the dissolved polymer into a solid fibre from a liquid chemical blend. Polymer solvent for spinning PPTA is generally 100per cent anhydrous (water free) sulfuric acid (H_2SO_4). The polymer is dissolved by mixing frozen sulfuric acid in powder form with the polymer in powder form and gently heating the mixture. This process, which differs from the more difficult DuPont process, was invented by Henri Lammers and patented by AKZO.

Industrial uses

Twaron is a para-aramid and is used in automotive, construction, sport, aerospace, military and industry applications, *e.g.*, “bullet-proof” body armor, fabric, and as an asbestos substitute.

- Protective gear (heat resistant/ ballistics): flame-resistant clothing, protective clothing and helmets, cut-fast or heat-hardy gloves, sporting goods, textiles, ballistic vests
- Composites: composite materials, technical paper, asbestos replacement, hot air filtration, sailcloth, speaker woofers, boat hull material, Fibre reinforced concrete, drumheads
- Automotive: brake pads, turbo hoses, V-belts and Timing belts, tires that incorporate Sulfron (sulfur modified Twaron), mechanical rubber goods reinforcement
- Linear tension: optical Fibre cables (OFC), ropes, wire ropes, cables, umbilical cables, electrical mechanical cable (EMC), reinforced thermoplastic pipes

KEVLAR

Kevlar is the registered trademark for a para-aramid synthetic Fibre, related to other aramids such as Nomex and Technora. Developed by Stephanie Kwolek at DuPont in 1965, this high-strength material was first commercially used in the early 1970s as a replacement for steel in racing tires. Typically it is spun into ropes or fabric sheets that can be used as such or as an ingredient in composite material components.

Currently, Kevlar has many applications, ranging from bicycle tires and racing sails to body armor because of its high tensile strength-to-weight ratio; by this measure it is 5 times stronger than steel. It is also used to make modern drumheads that withstand high impact. When used as a woven material, it is suitable for mooring lines and other underwater applications. A similar Fibre called Twaron with roughly the same chemical structure was developed by Akzo in the 1970s; commercial production started in 1986, and Twaron is now manufactured by Teijin.

Poly-paraphenylene terephthalamide – branded Kevlar – was invented by Polish-American chemist Stephanie Kwolek while working for DuPont, in anticipation of a gasoline shortage. In 1964, her group began searching for a new lightweight strong Fibre to use for light but strong tires. The polymers she had been working with at the time, poly-p-Phenylene-terephthalate and polybenzamide, formed liquid crystal while in solution, something unique to those polymers at the time.

The solution was "cloudy, opalescent upon being stirred, and of low viscosity" and usually was thrown away. However, Kwolek persuaded the technician, Charles Smullen, who ran the "spinneret", to test her solution, and was amazed to find that the Fibre did not break, unlike nylon. Her supervisor and her laboratory director understood the significance of her accidental discovery and a new field of polymer chemistry quickly arose. By 1971, modern Kevlar was introduced. However, Kwolek was not very involved in developing the applications of Kevlar.

Production

Kevlar is synthesized in solution from the monomers 1,4-phenylene-diamine (*para*-phenylenediamine) and terephthaloyl chloride in a condensation reaction yielding hydrochloric acid as a byproduct. The result has liquid-crystalline Behaviour, and mechanical drawing orients the polymer chains in the fiber's direction. Hexamethylphosphoramide (HMPA) was the solvent initially used for the polymerization, but for safety reasons, DuPont replaced it by a solution of *N*-methyl-pyrrolidone and calcium chloride. As this process had been patented by Akzo in the production of Twaron, a patent war ensued.

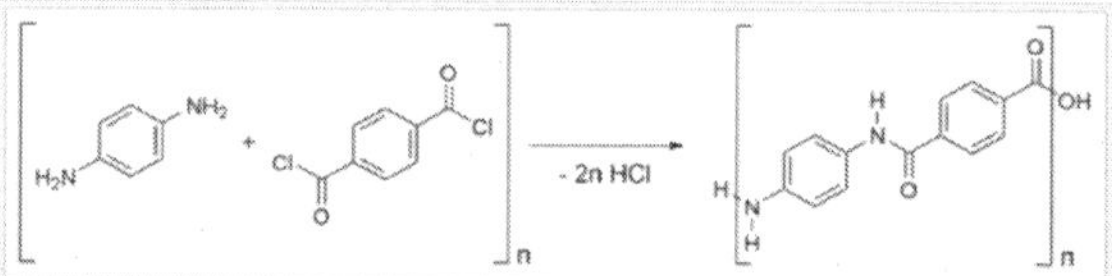

The reaction of 1,4-phenylene-diamine (*para*-phenylenediamine) with terephthaloyl chloride yielding kevlar

Kevlar (poly paraphenylene terephthalamide) production is expensive because of the difficulties arising from using concentrated sulfuric acid, needed to keep the water-insoluble polymer in solution during its synthesis and spinning.

Several grades of Kevlar are available:

1. *Kevlar K-29* – in industrial applications, such as cables, asbestos replacement, brake linings, and body/vehicle armor.
2. *Kevlar K49* – high modulus used in cable and rope products.
3. *Kevlar K100* – colored version of Kevlar
4. *Kevlar K119* – higher-elongation, flexible and more fatigue resistant
5. *Kevlar K129* – higher tenacity for ballistic applications
6. *Kevlar AP* – 15per cent higher tensile strength than K-29
7. *Kevlar XP* – lighter weight resin and KM2 plus Fibre combination
8. *Kevlar KM2* – enhanced ballistic resistance for armor applications

The ultraviolet component of sunlight degrades and decomposes Kevlar, a problem known as UV degradation, and so it is rarely used outdoors without protection against sunlight.

Structure and properties

When Kevlar is spun, the resulting Fibre has a tensile strength of about 3,620 MPa, and a relative density of 1.44. The polymer owes its high strength to the many inter-chain bonds. These inter-molecular hydrogen bonds form between the carbonyl groups and N*H* centers. Additional strength is derived from aromatic stacking interactions between adjacent strands. These interactions have a greater influence on Kevlar than the van der Waals interactions and chain length that typically influence the properties of other synthetic polymers and fibers such as Dyneema. The presence of salts and certain other impurities, especially calcium, could interfere with the strand interactions and care is taken to avoid inclusion in its production. Kevlar's structure consists of relatively rigid molecules which tend to form mostly planar sheet-like structures rather like silk protein.

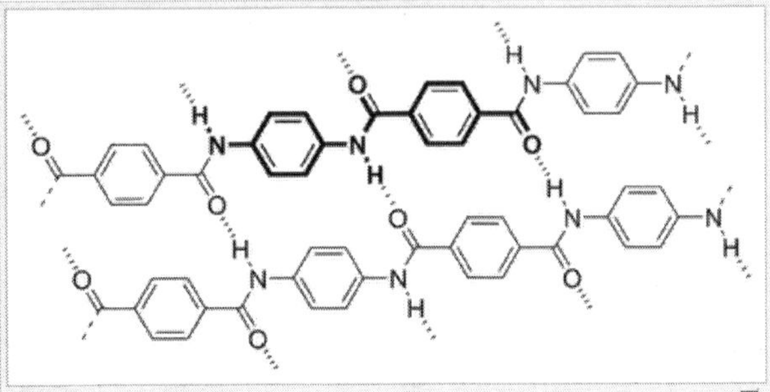

Molecular structure of Kevlar: **bold** represents a monomer unit, **dashed** lines indicate hydrogen bonds.

Thermal properties

Kevlar maintains its strength and resilience down to cryogenic temperatures ("196 °C); in fact, it is slightly stronger at low temperatures. At higher temperatures the tensile strength is immediately reduced by about 10–20per cent, and after some hours the strength progressively reduces further. For example at 160 °C (320 °F) about 10per cent reduction in strength occurs after 500 hours. At 260 °C (500 °F) 50per cent strength reduction occurs after 70 hours.

Applications

Protection

Cryogenics

Kevlar is often used in the field of cryogenics for its low thermal conductivity and high strength relative to other materials for suspension purposes. It is most often used to suspend a paramagnetic salt enclosure from a superconducting magnet mandrel in order to minimize any heat leaks to the paramagnetic material. It is also used as a thermal standoff or structural support where low heat leaks are desired.

Armor

Kevlar is a well-known component of personal armor such as combat helmets, ballistic face masks, and ballistic vests. The PASGT helmet and vest used by United States military forces since the 1980s both have Kevlar as a key component, as do their replacements. Other military uses include bulletproof facemasks used by sentries and spall liners used to protect the crews of armoured fighting vehicles. Even Nimitz-class aircraft carriers include Kevlar armor around vital spaces. Related civilian applications include Emergency Service's protection gear if it involves high heat (*e.g.*, tackling a fire), and Kevlar body armor such as vests for police officers, security, and SWAT.

Personal protection

Kevlar is used to manufacture gloves, sleeves, jackets, chaps and other articles of clothing designed to protect users from cuts, abrasions and heat. Kevlar based protective gear is often considerably lighter and thinner than equivalent gear made of more traditional materials.

Sports equipment

It is used as an inner lining for some bicycle tires to prevent punctures. In table tennis, plies of Kevlar are added to custom ply blades, or paddles, in order to increase bounce and reduce weight. It is used for motorcycle safety clothing, especially in the areas featuring padding such as shoulders and elbows.

- In kyudo or Japanese archery, it may be used as an alternative to more expensive hemp for bow strings. It is one of the main materials used for paraglider suspension lines.
- In fencing it is used in the protective jackets, breeches, plastrons and the bib of the masks.
- Tennis racquets are often strung with Kevlar.
- It is even used in sails for high performance racing boats.
- It is increasingly being used in the "peto", the padded covering which protects the picadors' horses in the bullring.

Shoes

With advancements in technology, Nike used Kevlar in shoes for the first time. It launched the Elite II Series, with enhancements to its earlier version of basketball shoes by using Kevlar in the anterior as well as the shoe laces. This was done to decrease the elasticity of the tip of the shoe in contrast to nylon used conventionally as Kevlar expanded by about 1per cent against nylon which expanded by about 30per cent. Shoes in this range included LeBron, HyperDunk and Zoom Kobe VII. However these shoes were launched at a price range much higher than average cost of basketball shoes.

It was also used as speed control patches for certain Soap Shoes models. and the laces for the adidas F50 adiZero Prime football boot.

Music

- Audio equipment: Kevlar has also been found to have useful acoustic properties for loudspeaker cones, specifically for bass and midrange drive units. Additionally, Kevlar has been used as a strength member in Fibre optic cables such as the ones used for audio data transmissions.
- Bowed string instruments: Kevlar can be used as an acoustic core

on bows for string instruments. Kevlar's physical properties provide strength, flexibility, and stability for the bow's user. To date, the only manufacturer of this type of bow is CodaBow. Kevlar is also presently used as a material for tailcords (aka tailpiece adjusters), which connect the tailpiece to the endpin of bowed string instruments.

- Drumheads: Kevlar is sometimes used as a material on marching snare drums. It allows for an extremely high amount of tension, resulting in a cleaner sound. There is usually a resin poured onto the Kevlar to make the head airtight, and a nylon top layer to provide a flat striking surface. This is one of the primary types of marching snare drum heads. Remo's "Falam Slam" Patch is made with Kevlar and is used to reinforce bass drum heads where the beater strikes.
- Woodwind reeds: Kevlar is used in the woodwind reeds of Fibracell. The material of these reeds is a composite of aerospace materials designed to duplicate the way nature constructs cane reed. Very stiff but sound absorbing Kevlar fibers are suspended in a lightweight resin formulation.

INDUSTRY OF SYNTHETIC FIBRES

There has been tremendous growth of synthetic fibre industry in India count Hilaire chardon was the first to make a fibre similar to silk. He made it from nitrocellu lose by dissolving it in ether. The first truly man-synthesized fibre, nylon was produced in 1938. All man-made fibres have some processes in common. They have been produced f.rom non-fibrous materials or if fibrous material were used originally, then they have lost their fibrous structure during processing. This is done by forcing the solutions through spinnerets.

These fibres are then permitted to harden during a specific. time and then are wound on bobbins or cones or they are deposited in "pots" as cakes of yarn. Man-made fibres are divided into two major classifications-non-thermoplastic and thermoplastic. The non-thermoplastic groups of man-made fibres have several origins, *e.g.,* cellulosic, alginates, minerals and protein base fibres.

Barring for the mineral. based fibres, these non-thermo-plastic fibres can be cared for as cotton, silk, wool whichever they similar most in their physical and chemical reactions. Heat will not melt them but they scorch easily at high temperatures. They are soft, absorbent, comfortable to wear, do not discharge static electricity and are mothproof. Like all other man-made fibres, under the microscope they appear as smooth rods, with black specks, as if they are delustered. None of these fibres can be positively identified in longitudinal sections by the microscope.

The first thermoplastic fibre to be made was acetate followed by nylon. Now apart from the acetates and nylons, there are acrylics, modacrylics, nefrils, olefins etc. These thermoplastic fibres react very fastly to heat. They soften and become pliable. This softening point varies with different fibres. When these fibres are at the softened stage, they can be shaped, pleated or embossed. These fibres, like other man-made fibres, must be stretched to orient the molecules in order to increase fibre strength, their tenacity, or to give them dimensional stability.

Nylon is generally heat set to stabilize the twist of yarn, to remove shrinkage, to increase wrinkle resistance, to set the grain in the fabrics, or to make washable pleats. These thermoplastic fibres have a number of properties in common. Their hygroscopicity is low.

Therefore they are uncomfortable in hot humid weather. They are washed easily and dry quickly. Water soluble stains are removed easily but not grease or oil. Majority of these fibres accommodate charges of static electricity. They are mostly wrinkle resistant. A microscopic examination reveals that they have the same longitudinal appearance as other man-made fibres. But their cross-sections vary and they can be identified with these.

VARIOUS FIBRES OF NON-THERMOPLASTIC NATURE

The Rayons

Rayon is an artificial, synthetic fibre made from cellulose.

Existence

The rayon fabrics were at first called 'artificial silks', as these resembled the natural silk in appearance. As recently as 1924, the name 'rayon' was provided to these fabrics. The rayons produced then were very lustrous, and, therefore, the name which means 'reflecting the sun's rays' is considered very suitable.

Although rayon fabrics have been commercially available for little more than fifty years, the idea of making such fabrics dated back to 1664 A. D. Robert Hook, an English Naturalist, prophesied in the year that a way would be found to make an artificial fibre which will similar natural silk. Then two centuries later, in 1884 Count Hilaire Chardonnet made a fibre which resembled silk, from nitrocellulose by dissolving it in ether, and produced it on a commercial scale.

Count Chardonnet is, therefore, considered the father of rayon fibre and the rayon industry. But the production of rayons did not make much development. In the startings of this century, U.K., Switzerland, Belgium, Austria, Germany and U.S.A. started making rayons. Since then much experimental and research work has been done to improve the process, as well as the quality of the yarn, in the course of which many processes were discovered and perfected.

These are called as :

(i) the viscose process,

(ii) the cuprammonium process

(iii) the cellulose acetate process.

Now all these processes are more in use than the one discovered by Chardonnet which is much more expensive. India has been importing rayon fabrics as well as rayon yarn. Since 1942, however, the Board of Scientific and Industrial Research has been making attempts for starting factories for making rayon yarn and many a plants have been set up since then. Bamboo, and Bagasse-cellulose from sugarcane a by-product of the sugar industry is used for making rayons in India.

Rayons Making Principles

Rayon, is made from cellulose by anyone of the former particular processes, but the important steps in each of these are:

(i) to treat cellulose chemically for making and rendering it a liquid,

(ii) to force the liquid through fine holes,

(iii) to change the liquid stream into solid cellulose filaments.

Process of Nitrocellulose

In this, cotton is reduced to nitrocellulose by treating cotton with sulphuric and nitric acids. This gives an inflammable material. This material is dissolved in ether and the fluid thus formed is compelled through tiny holes, into air. The ether evaporates and a fine thread is obtained, which is treated with sodium hydrosulphide to render it non-inflammable.

Process of Viscose

This process was discovered in 1892, but the yarn was commercially produced a few years later. In this process spruce. chips are utilised. These are first bleached, and then steeped in caustic soda to form alkali cellulose. It is treated with carbon bisulphide to form cellulose xanthate and is dissolved in dilute caustic soda solution. This is filtered and kept for ageing until a thick fluid is formed, which is known as viscose. This fluid is forced through fine jets into a coagulating. Solution of dilute sulphuric acid, which regenerates the cellulose into a continuous fibre.

This process is mainly used in the manufacture of rayons.

If a dull appearance is desired, a small proportion of a whiter opaque pigment (generally titanium dioxide) is added to the viscose solution.

Process of Cuprammonium

This process has been utilised since 1897. In this process, cotton linters are used. These are first boiled in soda and soda-ash, and then are bleached with chlorint. These are dissolved in a solution of copper sulphate and ammonium hydroxide. The liquid is left for ageing or ripening. This is then passed through fine jets into a solution of dilute acid when it is changed into regenerated cellulose filament. This filament is stretched to form a fine thread.

Process of Cellulose Acetate

This process has been commercially developed since 1918 although the process was discovered half a century earlier in 1869.

Rayon Yarn Spinning

The cellulose solution is passed through the spinrerets into a coagulating medium, in which the fibres get hardened. Many fibres are twisted together to form a yern. The yarn then is wound onspools; winding and rewinding is repeated using two spools and rewinding is repeated using two spools and each time giving a twist to the yarn. Then, finally, the yarn is would into skeins.

Rayon Its Features

Effect of Moisture and Friction

Regenerated rayons loose strength when wet but cellulose acetate is not as much affected in that respect by moisture. On drying, all of them regain theirstrength. Friction will weaken and spoil the lustre of the fibre, specially when wet.

Conduction of Heat

They are good conductors of heat.

Hygroscopic Moisture

Rayon fabrics absorb more moisture than cotton or linen but do not give out moisture as rerdily as those two fabrics.

Heat Effect

Heat affects them. Cellulose acetate rayon melts and fuses by the application of heat. The viscose and cuprammonium fibres are not greatly affected by heat, and the delustered effect given to these fabrics by heat can be diminished by the application of heat and moisture together as this will bring back the lustre.

Acids and Alkalies Action

All rayons are weakened even by dilute solution of acids. Cellulose acetate dissolves in acetate acid and formic acid Alkalies also have harmful effect on rayons.

Acetone Effect

Cellulose acetate dissolves in acetone, but not the other rayon.

Dyes Affinition

The viscose and cuprammomum fibres have great affinity for dyes used for cotton and linen, but the cellulose acetate rayon does not react to the same dyes. Rayons generally take dye-stuffs readily.

Bleaches

Action reducing bleaches may be used in cold, dilute solutions. Oxidizing bleaches will damage the fibre if these are not used with great care.

Rayon Lasting

Spun rayon made from tiny fibres is not so strong, but rayon fabrics made from long fibres are very strong, though net quite as durable as pure silk, cotton or linen. They are, however, generally more durable than weighted silk.

Elasticity

Rayon does not have natural elasticity. and, therefore, rayon clothes frail at the elbows, knees and the seams. Rayon requires greater allowance for seams. It is also important to have even tension and length of the stitches to avoid slipping. There are a number of non-thermoplastic fibres besides nylon, the foreign market. Some of them are avril, zantrel, qulin, Fortisan, Cornal, Tikel.

The processes utilised for the new fibres have not been published, but their preparation are known. Compared to regular rayons they have improved dimension stability, better wet stability and they have lower power of elongation.

Fibres of Thermoplastic Nature

Nylon

It was the fibre first to be synthesized from materials none of which had previously been fibrous in nature.

Origin. The history of nylon dates back to 1928, when Dr. Stine the Chemical Director of du Pont Company prevailed upon the Company to begin research work relating to new fibres. The research continued for some time and experiments were made on cellulose derivations, particularly the esters arid new types of esters, and also on certain nitrogen comprising derivatives of cellutose. These experiments were not successful Dr. Wallace H. Cenothen, at the same time made a study of poly-condensation whereby linear-polymers are produced, and this eventually led to the invention of nylon. In 1938, du Pont Company begen a plant for nylon production. This fibre was known as fibre 66. Later, the name nylon was given to it.

Nylon Making Process

Although it is often stated that nylon is made of coal, air and water, the actual synthesis is completely different. Nylon 66 is produced from an acid and a diamine, which are produced from coaltar derivaties. A mixture of two coal-tar products, a dibasic acid, adipic acid, and hexamethylene diamine containing nitrogen is heated to give a condensed product called nylon polymer. The, process is explained diagrammatically. Since each of the above compounds have six carbon atoms, the finished fibre made from this was known as Fibre 66.

The nylon polymer is me1ted and passed on chilled rollers when it comes out in the form of rolls, these are then cut into chips and stored. The chips are remelted and iltered through special filter packs. After filtering, the molten polymer is passed through the tiny holes of a metal disc called a spinneret and fine filaments are town. out into air where these get hardened.

These are then passed through a conditioner which moistens them so that a number of these filaments stick together to form a long thread which is wound on a reel. In order to increase the tensile strength and elasticity in the fibre, the filaments are drawn to about four times their original length by the application of force. In this operation tbe diameter of the fibre is reduced. Twisting of the yarn is also carried on at the same time by the use of suitable equipment for the purpose.

Nylon: Its Characteristics

It is a lustrous white transparent fibre and is both tough and pliable. Absorbency is low and this makes it a quick drying and easily washable fibre. Although it does not stain readily it tends to pick up colour, grease and soil if laundered with other garments. It is not affected by cold temperatures but looses strength and yellows at sustained high temperatures.

Pressing temperature of *24°F* is considered safe for nylon. It has good wrinkle resistance and crease recovery. It has very good abrasion resistance and is considered a very durable fibre. It is degraded by exposure to light. It is not chemically reactive to soap, alkalies or alcohol but reacts soon to acid.

Bleaches do not affect it and so are ineffectual for whitening discoloured nylon. They can be dyed and usually the colours are permanent. They are sometimes resin treated to stabilize the weave, silicone treated to improve softness or water-repellancy but because of their low absorbency they do not accept finishes readily.

Dacron

The work of Carothers paved the way for several other textile materials. Terrylene is one of these. Whinfield ard Dickson patented the method for making terrylene utilising teraphalic acid and ethylene glycol. This acid and alcohol are polymerized in vacuum at a high temperature and the polymer is extended in the shape of a riboon. The polymers (molten) are passd through spinnerettes with circular perforations. The Pont Company purchased patent rights of this fibre in the United States, and with some changes in processing, put Dacron as a new fibre on the market in 1953. Dacron and terrylene are essentially the same fibre.

The most significant properties of this fibre are its excellent resistance to wrinkling and creasing, both when dry and wet. It has good abrasion resistance, toughness, resilience, elasticity and stretch resistance.

Dacron has a property called as 'wicking'. Wicking is the ability of a fabric or fibre to pick up moisture and allow it to travel along the fibre topick up moisture and permit it to travel along the fibre, without actually absorbing it. Thus dacron is comfortable to wear inhot weather. Dacron melts and drops when exposed to fibre. It is self-extinguishing and does not flash burn. It has better resistance to weak acids and alkalies but is disintegrated by concentrated sulphuric acid. Dry cleaning solvents and bleaches do not effect this fibre. It is not easy to dye but can be satisfactorily coloured with acetate dyes.

Natural Proteins Based Materials

- *Ardil:* The raw material for the manufacture of ardil are the proteins, arachil and conarchin which can be obtained from ground. It is a consequence of the research carried by Imperial Chemical Industry (I.C.I.).
- *Fibrolane:* In principle the manufacture of fibrolane is resemble to that of ardil, the chief difference being that casein from milk is used instead of groundnut protein. People will be surprised if they are told that the suits they are wearing are of like instead of silk.

Mineral Fibres

These minerals are brought into use in textiles fibre making; asbestos a natural fibre; glass fibre and metallics a man-made fibre.

Asbestos Fibre

It is a naturally taking place mineral fibre. It is fibrous rock and is famous for its non-burning qualities. Under the microscope the fibres appear straight, smooth and needle-shaped. It does not burn or melt but short periods of high temperatures will reduce its strength. It is absorbent and has Yicking ability.

It is acid and alkali resistance and this makes it particularly desirable in the use of filters for chemicals and in industries where chemicals are utilised. Flameproof clothing for industrial and military purposes are also made from asbestos. It is used for all kinds of fire fighting equipment and insulation for steam pipes, brakes, and hundreds of other items where non-combustibility is inevitable.

Metallics Fibres

Metallics are defined as "a manufactured fibre composed of metal, plastic-coated metal, metal-coated plastic, or a core completely covered by metal." They are yarns and not fibres, but they are considered in fibre groupings as they replace fibres for textiles purposes. Gold and silver have long been utilised in India in textile fabrics. Aluminium and copper have a more recent origin. Metallics were used to add richness, glamour and glitte to fabrics.

Modern metallic fibres, with aluminium as the basic metal, are much cheaper, softer and lighter in weight. But because they are no more rich and luxurious and look cheap this may lead to their rejection.

Glass

'Fibre' is the trade mark for the glass fibre produced by the Owens-Corning Fiberglas Corporation of America and is die only Glass fibre about which information is available. Glass fibres are inorganic polymers based on silicon rather than carbon. These fibres have not become very popular as they are subject to severe abrasion and cut each other if they rub together.

They are used for curtains and drapery fabrics due to their attractive appearance, non-combustibility and are resistant to light deterioration. They are not suitable for wearing apparel as the ends of these fibres irritate the skin, they are heavy and absorbent. Much of the gtass fibre is used for insulation in buildings and for insulating electrical and other industrial equipment.

8

The Development of Fibers from Synthetic Polymers

Three events have created far-reaching changes in the historical pattern of consumption of textile fibers. In the Industrial Revolution of the eighteenth century cotton displaced wool as the world's principal fiber.

The commercial success of rayon and acetate in the first three decades of the twentieth century was the first incursion of man-made fibers into the traditional markets of cotton, gool, and other natural fibers. In this first achievement of the chemical industry artificial fibers were made from cellulose, a naturally occurring polymer.

Since the 1930's the textile world has been startled by an innovation whose horizon seems unlimited—the production of man-made fibers from synthetic polymers.

Although the new branch of the man-made fiber industry is still in its infancy, its future is secure enough to be regarded as an integral part of the raw-materials revolution of the twentieth century. In describing the advantageous properties of synthetics in contrast to natural fibers and rayon and acetate, textile chemists stress the striking and important differences which result from dissimilar moisture sensitivity.

Because of their outstanding functional properties the synthetics are assured a substantial portion of the fiber market of the future, one which will be attained in part at the expense of existing fibers.

The age of synthetics will lessen further man's dependence on nature for his supply of textile fibers and will diminish the importance of wool and cotton in agricultural economies.

In addition, the textile firm of the future will be less specialized than at present. The development of synthetic fibers has contributed to the current merger movement in the textile industry to obtain end-product diversity for survival in the new interfiber competition.

The regional impact of the new industry will find direct expression in the location of fiber plants and chemical plants.

CLASSIFICATION OF MAN-MADE FIBERS

From Natural Polymers

Viscose and acetate are the principal cellulosic fibers. Two other man-made fibers derived from natural polymers are also of commercial significance in the United States: glass fibers and Vicara.

From Synthetic Polymers

In this study the term "synthetic" fiber designates the second group of fibers only, those made from synthetic polymers. At the time research was initiated on this study the principal general-purpose synthetic textile fibers were nylon 66, Orlon, Acrilan, dynel, and Dacron; and this somewhat homogeneous group of five fibers was selected for the location analysis. Monofilament production, in general and synthetic fibers with specialized uses, such as saran, polyethylene, and polystyrene, lies outside the scope of this study.

MAN-MADE FIBER HISTORY

Although the story of the development of man-made fibers is a fascinating one, only its broad outline is presented here. The idea of man-made fibers is ancient and seems to have been inspired in part by a search for a substitute for silk, the ancient symbol of opulence.

The Development of Rayon and Acetate

Rayon came into existence through progress in the chemistry of cellulose and, circuitously, by way of the discovery of the electric lamp. In Edison's original electric lamp carbonized bamboo fibers were utilized for the filament but were found to be much too irregular for the purpose. To overcome this defect many substitute materials were examined and tested. In England Joseph W. Swan experimented with an electric lamp almost simultaneously with Edison. In 1883 Swan tried filaments made by squirting nitrocellulose solution through a die into a coagulating bath of water and alcohol. The potential use of these nitrocellulose filaments as textile materials was quickly discerned by Swan. Some fine threads were woven and exhibited as "artificial silk" at the Inventions Exhibition in 1885.

But Count Hilaire de Chardonnet was the inventor-entrepreneur who exploited the employment of nitrocellulose fibers for end uses in textile products. His own patents were first published in 1885, and under his direction factories for the production of nitro rayon were opened, first at Besançon in France and later in Belgium, England, Germany, Hungary, and elsewhere in France. Early nitro rayon was inflammable, and not until Swan developed a denitrating process in 1895 was cellulose nitrate made into a useful yarn. Even so the fiber was weak, especially when wet. Although the process flourished on a small scale, it was gradually superseded by cuprammonium and viscose rayon.

Cuprammonium Rayon

The first patents on cuprammonium rayon were published in France by Louis Despeissis in 1890. As early as 1856 Schweitzer had discovered that cellulose could be dissolved in solutions of cuprammonium hydroxide. German technicians Pauly, Fremery, Bronnert, and Urban made the vital improvements which led to commercial production in 1899, when the Vereinigte Glanzstoff Fabriken were formed. By 1909 cuprammonium rayon was a serious competitor to nitro rayon.

Viscose Rayon

Charles Stearn, a pioneer with Swan in the development of the carbon-filament electric lamp, led a small group of experimental workers to the production of the first skein of viscose rayon yarn in 1898. Stearn's work followed the discovery by C. F. Cross and E. J. Bevan in 1892 that alkali cellulose and carbon bisulphide react to form cellulose xanthate, a water-soluble ester. Stearn, together with Cross, formed the Viscose Syndicate, Ltd., in 1894 to extrude viscose filaments for carbon lamps and the Viscose Spinning Syndicate in 1898 to produce a textile yarn to compete with nitro rayon. The many difficulties encountered in perfecting viscose, however, were not overcome until 1910.

Cellulose Acetate

In 1894 another result of the fruitful inquiries of Cross and Bevan was a manufacturing process for the production of cellulose acetate, a chemical which had been discovered by Schutzenberger in 1865.

Through the work of Drs. Henry and Camille Dreyfus, the founders of the Celanese companies, cellulose acetate was successfully employed as a raw material for plastics and lacquer and in textile-fiber production. The first undertaking of the Dreyfus brothers was in 1911 at a factory in Basel, Switzerland, which manufactured cellulose acetate plastic (celluloid) and lacquer. During World War I operations were expanded to meet the demand for cellulose acetate for doping airplane wings; for this purpose factories were established in several Allied countries. At the end of the war surplus lacquer capacity was put to a new use-the first production of cellulose acetate yarn.

Growth of Rayon and Acetate Production

In the period from 1885 to 1920—from Chardonnet's nitrocellulose to cellulose acetate yarn—growth of output of the first manmade fibers was slow indeed. In 1920, 3,180.2 million pounds of textile fibers were consumed by mills in the United States, and the market share for rayon and acetate was less than one per cent. From the outset, rayon had an aesthetic appeal approaching silk and superior to cotton fabrics for intimate apparel. During the 1920's a sizable market developed for rayon and acetate in the underwear and dress fields. The early quality of rayon and acetate was not first-rate, and Heckert has noted that "it was not until the 1930's that the industry made a product comparing at all favorably with today's yarn." More rapid expansion of sales occurred from the 1930's on, especially after rayon and acetate were produced in staple form. Scientists turned out a rayon yarn of high strength and toughness in the same period, and it quickly replaced cotton as a material for tire cord. In 1956 total mill consumption of fibers in the United States was 6,485.7 million pounds, and the market for rayon and acetate had increased to 18.5 per cent.

The Development of Synthetic Fibers

The rapid development of the new synthetic fibers has been made possible by a number of factors. Perhaps the most important technical development is the fact that fiber synthesis now rests securely on basic advances which have been made in the chemistry of high polymers (a body of fundamental knowledge which also underlies the production of other synthetic materials, such as synthetic rubber, plastics, and finishes). Moreover, technical and market knowledge accumulated through the development of rayon and acetate was at work in paving the way for synthetic fibers. In addition, the significance of the coming of age of a textile science, in which the properties of fibers and fabrics can be measured in a precise way, cannot be overstressed.

Both the chemical and textile industries have been motivated by the necessity of novelty and variety in fibers and fabrics for the maintenance of old markets and the development of new. Interfiber price competition has also been instrumental in the search for new fibers, for the high prices of silk in the pre-World War II period and of wool in the postwar period have been a lure to the developers of nylon and the other new synthetics.

Advances in High-Polymer Chemistry

Through research in high-polymer chemistry it is now known that (in addition to chemical constitution) the fiber-forming properties of chemical compounds depend largely on their molecular weight, i.e., the size of the molecule formed during the reaction and the linear character of the molecule. Moreover, the molecules of many chemical compounds can combine with themselves to form giant, chainlike molecules via chemical change called polymerization. The molecular weight of the reaction product, a polymer, is many times that of the original compound. Thus, when subjected to polymerization under controlled conditions of heat and pressure, many chemical compounds can be converted into polymers of high molecular weight.

The superpolymers are unlike ordinary polymers, for an extraordinary transformation in physical properties occurs. Superpolymers produce filaments which can be permanently stretched or drawn after extrusion in a stretching operation (called orientation) which causes the molecules to be aligned in an orderly, lengthwise formation. The tensile strength, elasticity, and pliability of the resulting filament is vastly increased.

Before the 1920's there was no generally accepted theory of fiber-forming molecules. The theory that the behavior and properties of both natural and synthetic polymers is a consequence of a high molecular structure stems from the work, published in 1922, of the German chemist Staudinger, a 1954 Nobel prize winner. It seems clear that the new knowledge acquired in the chemistry of high polymers and the prior success of synthetic resins for plastics stimulated the research efforts to yield new manmade fibers.

Use of Vinyl Compounds

From vinyl compounds used initially as raw materials for plastics the first truly synthetic fibers were developed. Fremon has noted that I. G. Farbenindustrie began experimental production of a polyvinyl fiber in 1931. The early effort was abandoned in favor of "PeCe" fiber, made from polyvinyl chloride (after-chlorinated to render it soluble in acetone), which was offered for textile use in small quantities in 1934. In the United States Union Carbide pioneered in the development of synthetic yarns. From a special grade of Vinylite resin, a copolymer of vinyl chloride and vinyl acetate, synthetic fibers were made and sold under the trade name of "Vinyon" Vinyon fibers were first marketed in 1936. Although small-scale production of the first vinyl fibers was begun in 1939, both were handicapped by serious limitations for use as wearing-apparel fibers because of low softening points and excessive shrinkage at normal ironing temperatures.

The Development of Nylon 66

Nylon 66 was first evaluated by a number of hosiery manufacturers from yarn produced in a semi-works unit by a melt-spinning process which was also devised by Carothers. In October 1938 the decision was made to construct a full-scale plant at Seaford, Del. The original capacity, 4 million pounds, was doubled before construction of the first unit was completed. In retrospect, it is of interest to note that the 8-million-pound plant capacity for nylon within five years after Carother's laboratory discovery was almost equal to the 8.7 million pounds of rayon and acetate consumed by U. S. mills in 1920, thirty-six years after the first nitrocellulose filaments were made by Swan.

Other Nylon 66 Producers.

The second American concern to enter the nylon 66 market was the Chemstrand Corp., a joint venture of Monsanto Chemical and American Viscose.

Production under Du Pont license was begun in 1954 at a plant located at Pensacola, Fla.

Nylon 6

In Germany in 1938 another synthetic fiber of the polyamide class, nylon 6, was developed by Kleine and Schlach and produced under the name of Perlon L.

Continued Research in Polyvinyls

After the 1930's research in polyvinyls continued. The earliest U. S. patents on the fiber-forming properties of acrylonitrile were those of I. G. Farbenindustrie in 1938. Several American chemical companies including Du Pont, American Cyanamid, and Union Carbide and Carbon, were also at work during World War II on the use of acrylonitrile. The basic problem to be solved was the development of a solvent, and a whole series of patents were issued to Du Pont and other companies for this purpose during the period 1942-1946 until the problem was met by the use of dimethyl formamide.

Growth in the Market for Synthetic Fibers

Separate figures are not available on the consumption history of the new synthetic fibers; however, the astonishing growth of the new synthetics can be demonstrated by figures on plant capacity. In 1938 capacity for nylon 66 output was 8 million pounds. Synthetic fibers have found their way into the industrial market for use in filter cloths, electrical insulation, tire cord, chemical clothing, dye nets, automobile tops and seat covers, sewing thread, etc. In the realm of home furnishings the synthetics are used in draperies, upholstery, rugs, blankets, awnings, pillows, and other articles. But the major expansion has been in the apparel field for use in hosiery, in tricot knitted underwear and lingeries, in fabrics (alone or in blends with natural fibers or rayon and acetate), in linings, and in pile and fleece coatings.

LOCATION THEORY AND THE SYNTHETIC-FIBER INDUSTRY

An empirical location study must turn to location theory for a systematic coverage of all factors which might possibly influence the location of an industry and, in general, the geographical distribution of its plants. In this location study of the synthetic-fiber industry an attempt has been made to apply the general substitution approach to the locational equilibrium of the firm.

Underlying the location pattern of an industry are decisions made by business firms. The general area of economic theory which embraces this type of decision-making activity is a branch of microeconomics, the theory of the firm, and, in particular, the locational equilibrium of the firm. Traditionally, the theory of the firm has been made synonymous with price theory, the postulates of which are inferred through the implications of profit-maximizing behavior of the firm in three roles: first, in the process of production, i.e., the transformation of inputs into outputs; second, as a buyer of inputs; and third, as the seller of a product. Given the familiar data of the problem, including the structure of markets faced by the firm (defined in terms of numbers of buyers and sellers and homogeneity of product), profit-maximizing levels of output can be deduced. Price theory ignores the location problem because buying, production, and selling are looked upon as single-dimension activities, i.e., as occurring at a point.

This assumption is abandoned in the theory of the location equilibrium of the firm. As a buyer of inputs, the firm incurs costs in assembling them over space and is also confronted with regional differences in their availability; the impact of site on the technical possibilities of production is recognized; and the selling activity embraces a market area in which transportation costs are also incurred in delivering the product. Thus the attributes of plant-location theory stem from firm behavior in a fourth role as well: that of selecting from an array of geographical possibilities, minimum-cost production, and selling sites; and cost minimization is focused on variations which result from the factor of space.

The Weberian Framework

In his pioneering work Weber's signal contribution was the development of the principal categories of industrial location, largely on the basis of technical empirical knowledge. To Weber, the transportation network of a nation is the basic magnetic field of economic attraction which, through its lines of force, determines the position of industrial particles. He regarded transportation orientation as the basic principle of industrial location. Conceptually, competing magnetic fields are sometimes capable of causing particles to be drawn away from the transportation lines of force to their own. In short, all other types of orientation, such as power or labor, are deviations from minimum-cost transportation sites. Weber also singled out the importance of agglomeration economies in industrial location.

Although his types of location are suggestive for empirical studies, Weber's geometrical approach to the locational equilibrium of the firm, which, for example, requires the solution of location triangles, offers only a rough methodological guide.

The General Substitution Approach

That the Weberian location categories could be derived by means of the familiar substitution principle of price theory was suggested by Predohl in 1928, but it was not until 1949 that an operational framework was established for the general substitution approach to location theory by Isard. In his work Isard has extended the bounds of traditional production theory to include the locational equilibrium of the firm. Since cost minimization has a spatial dimension, it is specifically recognized by way of transport inputs. Within the enlarged framework the problem of production is that of choosing the right combination of capital, labor, land, and transport inputs.

From the most significant cost difference which occurs through variation over space the orientation of an industry can be determined. Isard distinguishes two types of substitution. The exploration of alternative outlays from the use of transport inputs can determine whether costs for a transport-oriented industry are minimized at market or at raw-material sites; and the traditional outlay substitutions of production theory between capital, labor, and land can be used to deduce other types of industrial location such as labor or power orientation.

APPLIED GENERAL SUBSTITUTION ANALYSIS AND THE SYNTHETIC-FIBER INDUSTRY

The general substitution approach to the locational equilibrium of the firm provides an effective operational tool for empirical analysis. This approach can be put in simple terms in the case in which a given market, or set of markets, is to be served by a plant with given physical inputs per unit of output and is able to draw raw materials from alternative supply areas. As this plant is moved from region to region, the effect on its costs and revenue via the substitution of inputs and of diverse outlays can be determined. Consider, for example, the alternative in the synthetic fiber industry of producing the chemicals for fiber synthesis at sites in chemical raw-material regions or of manufacturing them in fiber-market regions in the textile area of the United States.

Output at a raw-material site requires little outlay on movement of raw materials, but outlays are incurred on the transportation of the finished product and on utilities, labor, etc. At a market site a transportation cost on raw materials is substituted for the cost of moving the finished product, and the market-site utility, labor costs, etc., are substituted for those at the raw-material site. In addition, market-site agglomeration economies are substituted for those at the raw-material site. Substitutions such as these can be used to generate regional cost differences so that the most favorable region for plant location can be uncovered.

The Hypothesis

It is clear that the successful use of the general substitution approach in an empirical location study depends in a large measure upon the extent to which production functions, the physical relationship between various inputs (including transportation inputs) and output, can be obtained or estimated. If, for instance, the location of an ammonia plant is in question, it is necessary to know the raw-material inputs per unit of output, the utility and labor inputs, the manner in which these inputs vary with the size of plant, the manner in which the capital investment varies with size of plant, etc.

The hypothesis in this study of the synthetic-fiber industry is that:

- From an examination of the industry's production functions and market areas a spatial framework of regions which are suitable for plant location can be established.
- General substitution analysis can be used to convert the data of the production functions into production and distribution cost differences at selected sites.

In this manner basic regional advantages and disadvantages for plant location can be ascertained. The computation of regional differences in production and distribution costs will also point out the basic type of location orientation for each stage of the synthetic-fiber industry as well as the minimum-cost regions among all those tested.

COSTS AND LOCATION ANALYSIS

Although in the location analysis of the synthetic-fiber industry it might be desirable to employ all the costs which the chemical engineer or the cost accountant would include, from the standpoint of location theory the estimation of costs, hence the estimation of production functions, can be simplified. All inputs which have a locational significance must be identified, but the cost items which do not vary significantly from region to region can be set aside. Thus the application of the general substitution approach to the synthetic-fiber industry begins with prior determination of relevant cost categories to guide the estimation of production functions.

The basic cost categories in the theory of the firm are variable and fixed costs, and these are applicable in location analysis. With respect to the locational equilibrium of the firm, variable and fixed costs may vary spatially because of interregional differences in achievable agglomeration economies.

Cost Variation Plants of Equal Size

An examination of specific items in fixed production costs in both the chemical-intermediate and the fiber-producing stages of the synthetic-fiber industry leads to the conclusion that for any given predetermined size of plant regional uniformity in fixed charges can be assumed without doing violence to reality.

The chemical intermediates from which synthetic fibers are produced can be manufactured by more than one process, but the technology of petrochemical production is assumed in almost all cases in this study, since this particular route promises the most locational mobility for the future. In their study of the petrochemical industry Isard and Schooler found the assumption of regional uniformity of fixed costs for plants of given size to be a realistic one, after considering plant construction costs, interest charges, depreciation, plant maintenance, insurance costs, taxes, and land costs. In general, the analysis of the components of fixed charges also applies to the synthetic-fiber stage, even though fiber output is a combined chemical-textile operation.

In the past chemical-plant construction costs in warm-weather regions, such as Texas or California, have tended to be lower than in regions of the North by 10 or 12 per cent because certain structural parts unnecessary in the "outdoor" type of plant were thought to be indispensable in the "indoor" design for northern regions. However, Isard and Schooler note that the recent technical literature of chemical-plant construction points to the feasibility of the cheaper outdoor design for all regions. Plant construction cost inequalities which originate in labor-cost and material-cost variations cannot be systematically forecast and are not considered vital to the location problem in question.

For the large firms engaged in the development of synthetic fibers and their chemical intermediates ready access to the nation's capital market excludes the possibility of interest-cost disparities at different sites; therefore, the United States can be regarded as an area characterized by homogeneous capital availability for the location analysis. In considering the possibility of plant location in Puerto Rico, branch plants of American firms are assumed so that capital-cost uniformity applies to the island as well.

Since depreciation allowances are determined principally by tax legislation at the Federal level, there is little room for regional differences in this item of fixed costs. Plant maintenance has a regional variation and could be expected to be higher for cold weather regions, but the practice of performing major maintenance work in good weather limits the magnitude of the cost difference.

Insurance costs are significant in chemical production because of the risk hazard. Manufacture in a raw-material region with plants concentrated at specific sites raises plant insurance costs over areas in which chemical production is dispersed. Since the insurance cost differences are small and cannot be predicted in their variation with space, they are neglected in the cost analysis. To account for regional variations in the burden of taxes which enter into fixed charges is a difficult matter. Property taxes are important but are set by legislative bodies and are not amenable to location analysis. Some states offer property-tax exemptions, usually for a five-year period, as an inducement to the location of new industry.

The practice is so widespread, however, that its impact on location decisions tends in the long run to become neutralized regionally, although it may be a factor in the selection of a site within a region. For the cost analysis land outlays are assumed to be equal in all regions. The problem of minimizing land outlays in factory construction is not one of choosing the right region but the proper site within a region. Although interregional equality in fixed charges may be justifiably assumed, in both stages of the synthetic-fiber industry regional differences in variable costs do exist. Computation of the amounts of these differences requires first an estimation of the most important variable physical inputs per unit of output.

Agglomeration Economies

For the synthetic-fiber industry the major agglomeration economies are:

- Large-scale production economies;
- Integration economies from the output of several chemical products in a multiunit plant—savings are realized, for example, by the utilization of the by-product of one unit as a raw material for another;
- localization economies from the geographical concentration of chemical plants in a given area—a single large-scale chlorine plant may serve the needs of several users and make it unnecessary for each to construct a small-scale unit;
- External economies from the purchase of raw materials from large-scale producers-in Texas, for instance, large-scale oil refineries make available to chemical manufacturers huge streams of waste gas at low prices.

It may be valid to assume that for some products plants of the same predetermined size could be built in any region. In many cases this assumption would imply that agglomeration economies are uniform from region to region. However, if plants of the same predetermined size cannot be built in all regions, then location analysis must evaluate regional differences in variable and fixed costs caused by differences in the extent to which large-scale production and other agglomeration economies can be achieved.

All four of the above-mentioned types of agglomeration economy may be significant in the location of plants in the first stage of synthetic-fiber production, the manufacture of chemical intermediates. Since many of the intermediates have markets in other end uses, there may be regional differences in market potential. A site suitable for fiber manufacture may not draw the chemical stage unless it can serve a market for the excess production over the fiber requirement, for otherwise the chemical unit would suffer a large-scale production economy disadvantage. The other three types of agglomeration economy generally favor raw-material over market regions or, among market regions, those with the most highly developed chemical industries.

In this study agglomeration economies are evaluated for the chemical—intermediate stage through the estimation of the influence of size on fixed and variable costs. The other agglomeration economies-integration, localization, and external—are given qualitative consideration only. In the second stage of the industry, the manufacture of synthetic fibers, the main agglomeration consideration is large-scale production economies. The market for each synthetic fiber is spread out geographically over the entire textile area so that the output of a single plant is not sold entirely in a specific region. For this reason plant size is largely a function of technological and cost considerations common to all fiber producers and is somewhat independent of the region chosen for the location of a plant. Since the technological and cost considerations permit plants of economic size to be built in any region, agglomeration economies can be eliminated as a factor in the location of synthetic-fiber plants.

Conclusions of Preliminary Cost Analysis

For the first stage of the industry, the output of chemial intermediates, the assumption is made from the foregoing analysis that the location problem can be solved by comparing regional differences in variable costs alone for products for which the assumption is valid that plants of the same predetermined size can be built in any region. In most cases, however, the comparison of regional differences in variable costs must be augmented by the estimation of the effect on both fixed and variable costs of agglomeration economies.

With respect to the second stage, the manufacture of synthetic fibers from chemical intermediates, the assumption is made that the location problem can be solved by comparing regional differences in variable costs for plants of equal size.

FLAX

Flax is a bats Fibre used to manufacture linen textiles. It is derived from the stem of the annual plant, Linumusitatissimum, which grows in many temperate and subtropical areas of the world. The flax plant grows as high as 4 feet tall, and flax fibers are found below the plant surface held in place by woody material and cellular matter. The flax fibers are freed from the plant by a fermentation process called retting.

Chemical retting using acids and bases has been employed with some success; however, such processes tend to be more expensive than natural fermentation techniques. The fibers are then removed from the plant through breaking of the woody core, removal of the woody material, and combing. The resulting fibers are ready for spinning and are 12-15 inches (30-38 cm) in length.

Structural Properties

Flax is nearly pure cellulose and therefore exhibits many properties similar to cotton. Strands of individual flax fibers may consist of many individual Fibre cells—fibrils held together by a natural cellulosic adhesive material. The molecular chains in flax are extremely long, and the average molecular weight of molecules in flax is 3 million. There are swellings or nodes periodically along the Fibre which show up as characteristic cross markings. The cell walls of flax are thick, and the Fibre cross section is polygonal with a large lumen in the center.

Fig. Flax

Physical Properties

Flax is a strong Fibre, and the average tenacity of dry flax is 2-7 g/d; wet flax is 2.5-9 g/d. Like cotton, the wet flax Fibre is about 20per cent stronger than the dry Fibre. Flax has a low elongation at break; however, the Fibre is fairly elastic at very low elongations. Flax has a higher regain than cotton, and a 12per cent regain is observed at standard conditions. Flax possesses good heat conductivity and is cool to the touch. The heat and electrical properties of flax are similar to those of cotton, and flax possesses solubility characteristics identical to cotton. Flax is a highly rigid Fibre, and it tends to crease on bending due to its poor resiliency. The specific gravity of flax is 1.54, the same as cotton. Flax is a dull Fibre, but more lustrous 1inen fabrics can be obtained by pounding the linen with wooden hammers. Linen fabrics have good dimensional stability unless stresses have been introduced during the weaving process.

Chemical Properties

The chemical properties of flax are similar to those of cotton. Flax has good resistance to acids, bases, and chemical bleaches. Flax is resistant to insects and microorganisms, and only under severe moist warm conditions will it be attacked by mildews. Flax is only slowly degraded by sunlight, and it decomposes at temperatures similar to that observed for cotton.

End-Use Properties

Flax finds only 1imited use in modern textiles. The methods used to produce flax involve extensive Labour and cost; thus flax usage is reserved for special luxury applications. The strength of flax makes it superior to cotton in certain applications. Flax is resistant to abrasion, but like cotton suffers from lack of crease retention and poor wrinkle recovery. Linen fabric's superior absorbency, coupled with its cool crisp hand, contributes to its desirability as a prestige Fibre. The thermal properties and chemical resistance of flax make it suitable for many consumer applications. Linen can be laundered repeatedly without deterioration and ironed safely at temperatures as high as 220°C. Flax is very flammable. Flax is used in prestige items including handkerchiefs, towels, tablecloths, sheets, and certain garments.

CULTIVATION

Successful cultivation of cotton requires a long frost-free period, plenty of sunshine, and a moderate rainfall, usually from 600 to 1,200 mm (24 to 47 in). Soils usually need to be fairly heavy, although the level of nutrients does not need to be exceptional. In general, these conditions are met within the seasonally dry tropics and subtropics in the Northern and Southern hemispheres, but a large proportion of the cotton grown today is cultivated in areas with less rainfall that obtain the water from irrigation. Production of the crop for a given year usually starts soon after harvesting the preceding autumn. Planting time in spring in the Northern hemisphere varies from the beginning of February to the beginning of June. The area of the United States known as the South Plains is the largest contiguous cotton-growing region in the world. While dryland (non-irrigated) cotton is successfully grown in this region, consistent yields are only produced with heavy reliance on irrigation water drawn from the Ogallala Aquifer. Since cotton is somewhat salt and drought tolerant, this makes it an attractive crop for arid and semiarid regions. As water resources get tighter around the world, economies that rely on it face difficulties and conflict, as well as potential environmental problems. For example, improper cropping and irrigation practices have led to desertification in areas of Uzbekistan, where cotton is a major export. In the days of the Soviet Union, the Aral Sea was tapped for agricultural irrigation, largely of cotton, and now salination is widespread.

Cotton can also be cultivated to have colors other than the yellowish off-white typical of modern commercial cotton fibers. Naturally colored cotton can come in red, green, and several shades of brown.

GENETIC MODIFICATION

Genetically modified (GM) cotton was developed to reduce the heavy reliance on pesticides. The bacterium Bacillus thuringiensis (Bt) naturally produces a chemical harmful only to a small fraction of insects, most notably the larvae of moths and butterflies, beetles, and flies, and harmless to other forms of life. The gene coding for Bt toxin has been inserted into cotton, causing cotton, called Bt cotton, to produce this natural insecticide in its tissues. In many regions, the main pests in commercial cotton are lepidopteran larvae, which are killed by the Bt protein in the transgenic cotton they eat. This eliminates the need to use large amounts of broad-spectrum insecticides to kill lepidopteran pests (some of which have developed pyrethroid resistance). This spares natural insect predators in the farm ecology and further contributes to Non-insecticide pest management.

But cotton is ineffective against many cotton pests, however, such as plant bugs, stink bugs, and aphids; depending on circumstances it may still be desirable to use insecticides against these. A 2006 study done by Cornell researchers, the Center for Chinese Agricultural Policy and the Chinese Academy of Science on Bt cotton farming in China found that after seven years these secondary pests that were normally controlled by pesticide had increased, necessitating the use of pesticides at similar levels to non-Bt cotton and causing less profit for farmers because of the extra expense of GM seeds. However, a 2009 study by the Chinese Academy of Sciences, Stanford University and Rutgers University refuted this. They concluded that the GM cotton effectively controlled bollworm. The secondary pests were mostly miridae (plant bugs) whose increase was related to local temperature and rainfall and only continued to increase in half the villages studied. Moreover, the increase in insecticide use for the control of these secondary insects was far smaller than the reduction in total insecticide use due to Bt cotton adoption. A 2012 Chinese study concluded that Bt cotton halved the use of pesticides and doubled the level of ladybirds, lacewings and spiders. The International Service for the Acquisition of Agri-biotech Applications (ISAAA) said that, worldwide, GM cotton was planted on an area of 25 million hectares in 2011. This was 69per cent of the worldwide total area planted in cotton.

GM cotton acreage in India grew at a rapid rate, increasing from 50,000 hectares in 2002 to 10.6 million hectares in 2011. The total cotton area in India was 12.1 million hectares in 2011, so GM cotton was grown on 88per cent of the cotton area. This made India the country with the largest area of GM cotton in the world. A long-term study on the economic impacts of Bt cotton in India, published in the Journal PNAS in 2012, showed that Bt cotton has increased yields, profits, and living standards of smallholder farmers. The U.S. GM cotton crop was 4.0 million hectares in 2011 the second largest area in the world, the Chinese GM cotton crop was third largest by area with 3.9 million hectares and Pakistan had the fourth largest GM cotton crop area of 2.6 million hectares in 2011. The initial introduction of GM cotton proved to be a success in Australia – the yields were equivalent to the Non- transgenic varieties and the crop used much less pesticide to produce (85per cent reduction). The subsequent introduction of a second variety of GM cotton led to increases in GM cotton production until 95per cent of the Australian cotton crop was GM in 2009 making Australia the country with the fifth largest GM cotton crop in the world. Other GM cotton growing countries in 2011 were Argentina, Myanmar, Burkina Faso, Brazil, Mexico, Colombia, South Africa and Costa Rica.

Cotton has been genetically modified for resistance to glyphosate a broad-spectrum herbicide discovered by Monsanto which also sells some of the Bt cotton seeds to farmers. There are also a number of other cotton seed companies selling GM cotton around the world. About 62per cent of the GM cotton grown from 1996 to 2011 was insect resistant, 24per cent stacked product and 14per cent herbicide resistant.

Cotton has gossypol, a toxin that makes it inedible. However, scientists have silenced the gene that produces the toxin, making it a potential food crop.

ORGANIC PRODUCTION

Organic cotton is generally understood as cotton, from plants not genetically modified, that is certified to be grown without the use of any synthetic agricultural chemicals, such as fertilizers or pesticides. Its production also promotes and enhances biodiversity and biological cycles. United States cotton plantations are required to enforce the National Organic Programme (NOP). This institution determines the allowed practices for pest control, growing, fertilizing, and handling of organic crops. As of 2007, 265,517 bales of organic cotton were produced in 24 countries, and worldwide production was growing at a rate of more than 50per cent per year.

PESTS AND WEEDS

The cotton industry relies heavily on chemicals, such as herbicides, fertilizers and insecticides, although a very small number of farmers are moving Towards an organic model of production, and organic cotton products are now available for purchase at limited locations. These are popular for baby clothes and diapers. Under most definitions, organic products do not use genetic engineering. All natural cotton products are known to be both sustainable and hypoallergenic. Historically, in North America, one of the most economically destructive pests in cotton production has been the boll weevil. Due to the US Department of Agriculture's highly successful Boll Weevil Eradication Programme (BWEP), this pest has been eliminated from cotton in most of the United States. This Programme, along with the introduction of genetically engineered Bt cotton (which contains a bacterial gene that codes for a plant-produced protein that is toxic to a number of pests such as cotton bollworm and pink bollworm), has allowed a reduction in the use of synthetic insecticides.

Other significant global pests of cotton include the pink bollworm, Pectinophora gossypiella; the chili thrips, Scirtothrips dorsalis; the cotton seed bug, Oxycarenus hyalinipennis; the tarnish plant bug, Lygus lineolaris; and the fall armyworm, Spodoptera frugiperda, Xanthomonas citri subsp. malvacearum.

HARVESTING

Most cotton in the United States, Europe, and Australia is harvested mechanically, either by a cotton picker, a machine that removes the cotton from the boll without damaging the cotton plant, or by a cotton stripper, which strips the entire boll off the plant. Cotton strippers are used in regions where it is too windy to grow picker varieties of cotton, and usually after application of a chemical defoliant or the natural defoliation that occurs after a freeze. Cotton is a perennial crop in the tropics, and without defoliation or freezing, the plant will continue to grow. Cotton continues to be picked by hand in developing countries.

COMPETITION FROM SYNTHETIC FIBERS

The era of manufactured fibers began with the development of rayon in France in the 1890s. Rayon is derived from a natural cellulose and cannot be considered synthetic, but requires extensive processing in a manufacturing process, and led the less expensive replacement of more naturally derived materials. A succession of new synthetic fibers were introduced by the chemicals industry in the following decades. Acetate in Fibre form was developed in 1924. Nylon, the first Fibre synthesized entirely from petrochemicals, was introduced as a sewing thread by DuPont in 1936, followed by DuPont's acrylic in 1944. Some garments were created from fabrics based on these fibers, such as women's hosiery from nylon, but it was not until the introduction of polyester into the Fibre marketplace in the early 1950s that the market for cotton came under threat. The rapid uptake of polyester garments in the 1960s caused economic hardship in cotton-exporting economies, especially in Central American countries, such as Nicaragua, where cotton production had boomed tenfold between 1950 and 1965 with the advent of cheap chemical pesticides. Cotton production recovered in the 1970s, but crashed to pre-1960 levels in the early 1990s.

Beginning as a self-help Programme in the mid-1960s, the Cotton Research and Promotion Programme (CRPP) was organized by U.S. cotton producers in response to cotton's steady decline in market share. At that time, producers voted to set up a per-bale assessment system to fund the Programme, with built-in safeguards to protect their investments. With the passage of the Cotton Research and Promotion Act of 1966, the Programme joined forces and began battling synthetic competitors and re-establishing markets for cotton. Today, the success of this Programme has made cotton the best-selling Fibre in the U.S. and one of the best-selling fibers in the world.

Administered by the Cotton Board and conducted by Cotton Incorporated, the CRPP works to greatly increase the demand for and profitability of cotton through various research and promotion activities. It is funded by U.S. cotton producers and importers.

USES

Cotton is used to make a number of textile products. These include terrycloth for highly absorbent bath towels and robes; denim for blue jeans; cambric, popularly used in the manufacture of blue work shirts (from which we get the term "blue-collar"); and corduroy, seersucker, and cotton twill. Socks, underwear, and most T-shirts are made from cotton. Bed sheets often are made from cotton. Cotton also is used to make yarn used in crochet and knitting. Fabric also can be made from recycled or recovered cotton that otherwise would be thrown away during the spinning, weaving, or cutting process. While many fabrics are made completely of cotton, some materials blend cotton with other fibers, including rayon and synthetic fibers such as polyester. It can either be used in knitted or woven fabrics, as it can be blended with elastine to make a stretchier thread for knitted fabrics, and apparel such as stretch jeans.

In addition to the textile industry, cotton is used in fishing nets, coffee filters, tents, explosives manufacture (see nitrocellulose), cotton paper, and in bookbinding. The first Chinese paper was made of cotton Fibre. Fire hoses were once made of cotton.

The cottonseed which remains after the cotton is ginned is used to produce cottonseed oil, which, after refining, can be consumed by humans like any other vegetable oil. The cottonseed meal that is left generally is fed to ruminant livestock; the gossypol remaining in the meal is toxic to monogastric animals. Cottonseed hulls can be added to dairy cattle rations for roughage. During the American slavery period, cotton root bark was used in folk remedies as an abortifacient, that is, to induce a miscarriage.

Gossypol was one of the many substances found in all parts of the cotton plant and it was described by the scientists as 'poisonous pigment'. It also appears to inhibit the development of sperm or even restrict the mobility of the sperm. Also, it is thought to interfere with the menstrual cycle by restricting the release of certain hormones.

Cotton linters are fine, silky fibers which adhere to the seeds of the cotton plant after ginning. These curly fibers typically are less than D_8 inch (3.2 mm) long. The term also may apply to the longer textile Fibre staple lint as well as the shorter fuzzy fibers from some upland species. Linters are traditionally used in the manufacture of paper and as a raw material in the manufacture of cellulose. In the UK, linters are referred to as "cotton wool". This can also be a refined product (absorbent cotton in U.S. usage) which has medical, cosmetic and many other practical uses. The first medical use of cotton wool was by Dr. Joseph Sampson Gamgee at the Queen's Hospital (later the General Hospital) in Birmingham, England.

Shiny cotton is a processed version of the Fibre that can be made into cloth resembling satin for shirts and suits. However, it is hydrophobic (does not absorb water easily), which makes it unfit for use in bath and dish towels (although examples of these made from shiny cotton are seen).

The name Egyptian cotton is broadly associated with quality products, however only a small percentage of Egyptian cotton production is actually of superior quality. Most products bearing the name are not made with the finest cottons from Egypt. Pima cotton is often compared to Egyptian cotton, as both are used in high quality bed sheets and other cotton products. It is even considered superior by some authorities. Pima cotton is grown in the American southwest.

INTERNATIONAL TRADE

The largest producers of cotton, currently (2009), are China and India, with annual production of about 34 million bales and 27 million bales, respectively; most of this production is consumed by their respective textile industries. The largest exporters of raw cotton are the United States, with sales of $4.9 billion, and Africa, with sales of $2.1 billion.

The total international trade is estimated to be $12 billion. Africa's share of the cotton trade has doubled since 1980. Neither area has a significant domestic textile industry, textile manufacturing having moved to developing nations in Eastern and South Asia such as India and China. In Africa, cotton is grown by numerous small holders. Dunavant Enterprises, based in Memphis, Tennessee, is the leading cotton broker in Africa, with hundreds of purchasing agents. It operates cotton gins in Uganda, Mozambique, and Zambia. In Zambia, it often offers loans for seed and expenses to the 180,000 small farmers who grow cotton for it, as well as advice on farming methods. Cargill also purchases cotton in Africa for export.

The 25,000 cotton growers in the United States of America are heavily subsidized at the rate of $2 billion per year although China now provides the highest overall level of cotton sector support. The future of these subsidies is uncertain and has led to anticipatory expansion of cotton brokers' operations in Africa. Dunavant expanded in Africa by buying out local operations. This is only possible in former British colonies and Mozambique; former French colonies continue to maintain tight monopolies, inherited from their former colonialist masters, on cotton purchases at low fixed prices.

LEADING PRODUCER COUNTRIES

The five leading exporters of cotton in 2011 are (1) the United States, (2) India, (3) Brazil, (4) Australia, and (5) Uzbekistan. The largest Non-producing importers are Korea, Taiwan, Russia,, and Japan.In India, the states of Maharashtra (26.63per cent), Gujarat (17.96per cent) and Andhra Pradesh (13.75per cent) and also Madhya Pradesh are the leading cotton producing states, these states have a predominantly tropical wet and dry climate. In Pakistan, cotton is grown predominantly in the provinces of Punjab, and Sindh. The leading area of cotton production is the south Punjab, comprising the areas around Rahim Yar Khan, Bahawalpur, Bahawalnagar, Multan, Dera Ghazi Khan, Muzaffargarh, Vehari, and Khanewal. In Sindh Sanghar is the most important cotton producing district. Faisalabad is a leader in textiles within Pakistan. Punjab has a tropical wet and dry climate throughout the year therefore enhancing the growth of cotton. In the United States, the state of Texas led in total production as of 2004, while the state of California had the highest yield per acre.

FAIR TRADE

Cotton is an enormously important commodity throughout the world. However, many farmers in developing countries receive a low price for their produce, or find it difficult to compete with developed countries. This has led to an international dispute (see United States – Brazil cotton dispute):

- On 27 September 2002, Brazil requested consultations with the US regarding prohibited and actionable subsidies provided to US producers, users and/or exporters of upland cotton, as well as legislation, regulations, statutory instruments and amendments thereto providing such subsidies (including export credits), grants, and any other assistance to the US producers, users and exporters of upland cotton.
- On 8 September 2004, the Panel Report recommended that the United States "withdraw" export credit guarantees and payments to domestic users and exporters, and "take appropriate steps to remove the adverse effects or withdraw" the mandatory price-contingent subsidy measures.

While Brazil was fighting the US through the WTO's Dispute Settlement Mechanism against a heavily subsidized cotton industry, a group of four least-developed African countries — Benin, Burkina Faso, Chad, and Mali — also known as "Cotton-4" have been the leading protagonist for the reduction of US cotton subsidies through negotiations. The four introduced a "Sectoral Initiative in Favour of Cotton", presented by Burkina Faso's President Blaise Compaoré during the Trade Negotiations Committee on 10 June 2003. In addition to concerns over subsidies, the cotton industries of some countries are criticized for employing child Labour and damaging workers' health by exposure to pesticides used in production. The Environmental Justice Foundation has campaigned against the prevalent use of forced child and adult Labour in cotton production in Uzbekistan, the world's third largest cotton exporter. The international production and trade situation has led to "fair trade" cotton clothing and footwear, joining a rapidly growing market for organic clothing, fair fashion or "ethical fashion". The fair trade system was initiated in 2005 with producers from Cameroon, Mali and Senegal.

9

Cotton Textile

The textile and apparel sectors are the largest generators of unskilled manufacturing employment in the world. Light and medium manufacturing industries, such as textiles and apparel, have the capacity to not only lift large segments of the population from poverty, but also to employ some of the most disadvantaged members of society. It is, therefore, crucial that Mozambique consider the potential viability and promotion of these sectors.

The purpose of this study is to provide technical input on the main constraints and opportunities for an integrated national industrial strategy for the development of the cotton based textile value chain. The main links in the cotton based textile value chain to be considered are:

- Cotton (including seed cotton, ginneries and lint);
- Textile manufactures (including spinning, fabric forming, dying and finishing);
- Apparel production.

Each link in the national supply chain has the potential to develop on its own merits, however, the possibility of supply chain integration must be explored and pursued when it offers the chance for stakeholders to improve their mutual competitive advantage and to share in increased value creation. Competitive advantages from a national integration strategy may arise from:

- Reducing or eliminating international transport costs and hence retaining value added within the national supply chain;
- Improving control over supply chain links to align quality with end customer requirements;
- Reducing lead times and improving responsiveness of the supply chain;
- Improving market access for products under preferential trade agreements.

This model of industry integration is based on market led supply chain linkages and the creation of value within the supply chain to the betterment of all stakeholders. As such, it represents a more complex process then colonial supply chain integration whose focus was the transfer of value from one stakeholder to another with little value creation. In a value chain approach, integration of the industry has to better all stakeholders through the creation of value. This is an approach that is, today, at the core of global supply chains—producers focus on what they do best, and cooperate where there is an opportunity for value creation. Global producer's integrate and\or divest themselves based on these criteria.

VALUE CHAIN APPROACH

In a value chain approach, constraints to industry growth can have a two fold impact on the supply chain. First, a constraint can limit the growth of an individual link, for example, labour laws may limit the growth of a downstream garment industry but not be directly important to production in the cotton and spinning sectors. However, the growth of the garment industry indirectly limits the opportunities for a textile industry to develop, and this reduces the opportunities of cotton growers and ginners to sell their cotton directly to large spinners, knitters and weavers. At the same time, the lack of major transportation arteries might not directly affect a textile or making up industry (presumably located near ports) but it could eliminate the opportunity to integrate into the rural cotton sector, thereby missing a chance to purchase local lint without incurring the full cost of transportation from international logistics agents, and reducing shipping delays. With cotton grown in the North of the Mozambique and the best access to infrastructure and international shipping routes in the South of Mozambique, this challenge requires special consideration in an industrial integration strategy.

The goal of this strategy is to report illuminate key constraints to investment and growth in individual sectors and identify areas where the constraints for value chain integration exist. The elements of value chain integration and value chain growth are complex, and are often only truly known when investors engage in the process of production and investment. Therefore, the constraints in this report should be considered as guide posts for industry development. Consequently, it is of the utmost importance that the government of Mozambique actively engages investors to support them in their exploration of Mozambique's cotton value chain. The government must address constraints that are overwhelming to investors, and continue to mitigate any constraints that remain, providing the investor the greatest opportunity to locate and succeed in Mozambique. It would be dangerous to assume some natural comparative advantage alone, such as low wages or raw materials, could be compelling enough to entice investors. Competitive advantages rarely reveal themselves so easily, and almost never without the trial and error of investors and market discipline.

The report is organized into four major sections. A profile of Mozambique's major value chain links, including cotton, textile and apparel production. The internal constraints and opportunities to investment and growth of the individual segments and to the integration of the value chain. The external constraints and opportunities for growth and investment, these factors are often categorized as market access issues as defined by international trade agreements. Finally, a proposal for a national action plan is presented including interventions and goals.

COTTON TEXTILE SECTOR

Cotton has been cultivated in Mozambique for over 50 years. In the years just before independence, Mozambique claimed one the largest vertically integrated textile and apparel industries in Africa, spanning the entire supply chain from fiber to spinning and fabric forming to garment production. The period directly after independence saw a dramatic decline in cotton production and the virtual disappearance of the downstream textile and apparel industries. Today Mozambique's seed and lint cotton sectors are about to exceed peak production levels achieved in the pre-independence era. The downstream textile and apparel sectors are paused to make a fresh start in the now global industry encouraged by preferential access to major markets. The following paragraphs provide an overview of the three sub-sectors, cotton, textile and apparel from a statistical, geographical and structural perspective.

SEED AND LINT COTTON SUB—SECTOR

The cotton fiber sub-sector encompasses two major stages including the cultivation of seed cotton and the ginning of seed cotton into lint which can be sold to world markets. Over 75 percent of current production is in the Northern provinces of Nampula and Cabo Delgado. Small farm holdings make up 98 percent of cotton production with the balance of production consisting of private land holding companies and joint venture (JV) firms. Illustrates raw seed cotton output for the years 2003 through 2005 compared to the peak production level achieved in 1973. Cotton production has consistently risen for three years in a row and is projected to reach 140,000 tons in 2006. Even still, experts estimate Mozambique's full potential to produce seed cotton to be in excess of 400,000 tons per annum.

Major actors in the Mozambique cotton sector include the Mozambique Cotton Institute, part of the Ministry of Agriculture and Rural Development.

The cotton staple fiber produced is determined largely by the length of the growing season and regional precipitation levels. The cotton fiber grown in Mozambique is classified as "middling to middling" and is among the longest African cotton fiber produced, outside of Egypt. Average fiber length is 2.86 cm with a minimum length of 2.77 cm1 or about 1 1/8th inch. This long staple commands a price premium of between 10-20 percent over shorter staple fiber in world markets, if it is supplied free of contaminants and if it is graded properly.

When fiber of this length is properly cleaned and graded it can be used to make an exceptionally wide array of yarns ranging from 5 to 50 counts or even 60 count yarns with exceptional quality standards enforced. Yarns in this broad range can be used to produce everything from fashionable "ring spun" denim jeans2 (5—10 count yarns) to t-shirts and fashion forward knit and woven shirts. On the home textile front, yarns of these grades could be used in mid to higher end bed sheeting and fine towels.

Although cotton grown in Mozambique is suitable for a wide range of products, Mozambique cotton is ideal for ring spun 100 percent cotton yarns of 30-36 count resulting in a competitive and sought after product. Yarns of this count could be utilized in a wide range of knit products from t-shirts and underwear to knit shirts and sweaters.

Joint ventures between the Mozambique government and international investors have been established to manage the conversion and sale of seed cotton into lint cotton. The JVs were established in the form of concessions whereby local farmers are required to sell their seed cotton harvest to geographically defined ginners. In return, the JVs provide critical financing and outreach programs for small farmers. The prices for seed cotton are set by government agents to insure small farmers a minimum price level and hence stability of income.

With seventeen ginneries currently operating with a capacity to process approximately 200,000 tons of cotton per annum, there is significant room for expanding the production of cotton lint. Investments in newer ginning machinery continue to grow and it is hoped with sustained harvests, the upgrading of equipment will continue. Lacking a spinning industry, Mozambique exports 100 percent of cotton lint produced, principally through the JVs and international cotton traders. There are few direct linkages to end customers.

The future of cotton production in Mozambique is promising. New ventures in the area of organic cotton cultivation have not yet been proven in Mozambique, but it is expected that small farm holdings will embrace this growing market opportunity in the near future.

Textiles Sub-Sector

After independence in 1975, Mozambique's nascent textile industry sprung into action with six major textile firms engaged in the key activities encompassing the entire textile supply chain: spinning, weaving and knitting, dying and finishing of fabric. Not all of these operations depended exclusively on cotton fiber, since some were located in the South of the country, far from the major cotton growing areas in the North. Instead, the Southern mills served the local markets with low cost polyester fabrics. Northern mills, favored by their location to local cotton crops, produced cotton uniforms and cotton blankets from second and third grade cotton, again, primarily for local markets.

In the early 1990s, hampered by poor market opportunities, low capacity utilization and new laws, textile firms began to falter until the final textile mill closed in 2001. Significant textile machinery remains in Mozambique in various states of repair. Most equipment is over ten years old and include over 30,0000 spindles and possibly 100 shuttless looms that may be revived for producing for the local market. The government of Mozambique, in cooperation with its private owners, is seeking investors to revitalize and renew this machinery for textile manufacturing. However, efforts to divest of these buildings and machinery have been hampered by significant financial liabilities, the age of the equipment and legal entanglements of the firms that formally employed them. New investors may find green field opportunities attractive, especially when considering that old textile mills and equipment may be located in remote areas of the country without significant regard for access to seed or lint cotton, international shipping access, downstream users and ports. It is likely new investors will want to explore new locations to take advantage of Mozambique's natural and geographical advantages while balancing access to end using industries and transportation.

Apparel Sub—Sector

Mozambique's apparel industry is currently comprised of one medium sized factory employing about 400 workers in the cutting and sewing of specialty uniforms for the South African, US and EU markets. This firm has been in operation since 2003, when South African investors purchased a formerly state owned company. A visit to local markets reveals numerous local tailors and a significant micro-enterprise based stitching industry. Pre-production capabilities, such as pattern and marker making are absent. With significant market access benefits from the US African Growth and Opportunity Act (AGOA), the South African Development Community (SADC) and the EU, Mozambique's apparel industry holds distinct advantages over competitors in Asia.

SECTOR SPECIFIC ANALYSIS OF COTTON

SEED AND LINT COTTON SUB—SECTOR

Cotton is a widely traded commodity with the largest consuming markets being the EU, India, Pakistan, China and Brazil. Tariffs on raw cotton imports, not carded or combed are low—at 5.0% percent of less. Some notable restraints do exist, including the US, which applies specific duties and quotas on imports of raw cotton. The US does not provide preferential access for cotton under AGOA. Under the EU ACP*Cotonou* program and the EU EBA program for least developed countries, Mozambique is eligible for tariff free for treatment seed and lint cotton. However, the EU does reserve the right to limit imports under these programs if they are causing or threaten to cause market disruptions. The Southern African Development Community (SADC) provides Mozambique preferential access to the South African market for seed and lint cotton. South Africa applies a specific duty of 1.6 Rand per KG or the estimated tariff equivalent of 15%. Mozambique is exempt from this tariff; since cotton is, an agricultural product which is wholly formed in Mozambique, meeting the rule of origin.

It is important to note that organic cotton certifications are not globally recognized, so each importing country may have a separate set of rules for deciding which cotton qualifies for this certification. Technical grading of cotton is a significant issue, and is the focus of efforts of the International Cotton Advisory Committee (ITAC).

Textiles Sub—Sector

Unlike cotton, an agricultural product that does not confront complex rules of origin, textiles are manufactured and must conform to widely varied rules of origin to benefit from any preferential access programs. For the purposes of market access analysis, there are three products\proceses that must be considered:

- Yarns (spinning);
- Fabrics (knit and woven);
- Home and industrial products (made-ups).

Each category is potentially subject to different rules and tariffs regarding market access provisions.

In general, rules of origin for textiles are based on a transformation system. With the transformation based system, certain processes must be carried out on non-originating materials or inputs before originating status can be conferred. In the case of textiles, there are three major processes that take place: spinning, fabric forming, and finishing. Many countries do not recognize finishing as a significant process, meaning two processes are left to confer origin, spinning and fabric forming. Since there are no significant manufacturing processes before yarn spinning, yarn must be spun from local fibers to be considered originating and benefit from most programs. Fabric must be made from yarn spinning and fabric forming, but the fiber can be from anywhere8. This is in contrast to a value added system, which counts the value added to materials as the criteria for certifying goods as originating. There are important difference between the US and EU preferential programs and their rules of origin.

EU –ACP\EBA

The EU ACP and EBA agreements provide Mozambique duty free access for products meeting the EU rules of origin. With yarn tariffs averaging six percent and fabric tariffs of approximately 8 percent, preferential access is important.

To benefit from the relief of these duties, yarns must be spun of local fiber and fabrics must be made from yarn spun in the beneficiary country or region. In the case of the EBA agreement, all processes, including yarn spinning and fabric forming must occur in Mozambique. In the case of the ACP agreement (which expires in 2008), Mozambique may cumulate origin with other ACP countries. This means that yarns spun in Mozambique and formed into fabric in another ACP country (Mauritius for example) will be eligible for the elimination of tariffs, at least as long as the ACP agreement is effective.

Southern African Development Community (SADC)

The SADC free trade has eliminated tariffs on South African imports of yarns and fabrics from Mozambique. Today, South African most favored nation tariffs average 15 percent for yarns and 20 percent for fabrics. Like the EU, SADC rules require two stages of transformation for fabrics to benefit from tariff elimination. SADC does not support the cumulating of processes across SADC countries. So, fabrics must be made from fibers spun in a single beneficiary country. Yarns must be spun from local fibers to be originating—a major challenge for many regional producers and a potential advantage for a producer located in Mozambique.

African Growth, Opportunity and Investment Incentive Act (AGOA)

Early AGOA agreements excluded all but folk art textiles from preferential treatment. The new AGOA Investment Incentive act, signed in December of 2006, changed the rules to include all textiles and textile products (listed earlier) as long as these products are wholly formed in lesser developed SSA countries. With tariffs that range from 8 percent on yarns up to 12 percent on certain home textile products the AGOA provides significant preferences if the rules of origin can be met.

Apparel—sub-Sector

Like textile products, apparel products exported from Mozambique can potentially benefit from tariff relief in major markets, subject to rules of origin that define products that benefit from preferential programs. The EU currently offers two potential programs, with different rules of origin, the ACP agreement and the EBA program. The special rules of origin for apparel under the US AGOA arrangement, called the third party fabric provision have recently been extended through 2012 – a six year extension. South Africa offers complete relief from tariffs under two provisions in the SADC free trade agreement: one provision provides a quota for products with a similar rule of origin as the AGOA third party fabric provision; the second is unlimited tariff relief for products meeting stricter rules.

Since its inception in 2000, the US AGOA program has grown to be the most significant contributor to growth in the SSA apparel industries. SSA countries such as Lesotho, South Africa, Kenya, Madagascar and Mauritius now credit AGOA with the creation of hundreds of millions of dollars in apparel exports and tens of thousands of sewing and related jobs. No other industry in the agriculture, mining or manufacturing sectors has benefited as much from AGOA as the Apparel industry.

The first key to the apparel industries success under AGOA has been relief from high US apparel tariffs, averaging between 16 and 32 percent10. Generally, products made of principally cotton yarns and fibers draw average duties of between 16 and 22 percent, with the highest US apparel tariffs being applied to products made of man-made-fibers. Still, the margin of preference on cotton products is significant. To some degree, it is a significant counter balance to Mozambique's distance from the US market.

The third party fabric provision, the second key to AGOA success, provides easy adaptation to established supply chains with strong ties to Asia. The AGOA Investment Incentive Act extends the third party fabric provision through September 2012. Importantly, The AGOA Investment Incentive Act provides for new limits on the use of third party fabrics, especially apparel made of denim. As a general rule, any fabric deemed to be in abundant supply by the US government will be excluded from the third party fabric provision, or have its use limited. Any producer seeking to utilize third country fabrics for apparel exports to the US will have to monitor the US government assessments of fabrics in deemed to be in abundant supply in SSA.

Apparel constructed of fabrics formed in SSA with SSA formed yarns is not subject to the limits set by the abundant supply provision or the 2012 limit on the use of third party fabrics. Instead, the AGOA provides for tariff free exports of these products through 2015, subject to a cap, which has never been filled.

Southern African Development Community (SADC)

The Southern African Development Community is comprised of xx countries. SADC has an established history of free trade in apparel among its partners, in particular with South Africa, the largest and most affluent market for apparel in the region. With South African tariffs on apparel from non-SADC countries at 40 percent, tariff free access provides a major advantage to apparel producers seeking to sell into the South African market.

SADC apparel rules of origin require two transformations (reviewed earlier), which means that cutting and sewing of fabric is not sufficient to secure tariff benefits. Instead, the fabric must be form (knit or woven) in Mozambique as well as cut and sewn to be eligible for tariff free treatment. This rule is the same for all SADC countries. In contrast to the US AGOA, the SADC rule of origin does not permit the cumulating of processes across SADC countries, so fabric that is formed in South Africa, then cut and sewn in Mozambique is not eligible for tariff free treatment.

Similar to the US AGOA program, South Africa provides for a limited amount of apparel made from third party fabrics to benefit from complete tariff relief. The program that provides these benefits is called the MMTZ and is offered by South Africa to the least developed countries of SADC. Importantly, MMTZ imports into South Africa are limited by quota. The MMTZ quota has been reduced since its inception, but current quota levels are not fully utilized by Mozambique. Exporters seeking to use this provision should contact the South African Revenue Service (SARS) to monitor MMTZ quota levels and their future evaluation.

EU-Lome and Everything but Arms Arrangement

The EU provides tariff relief to developing countries under several different programs. Mozambique, classified as a least developed country is eligible for complete tariff relief under two arrangements, the EU *Lome* or African Caribbean and Pacific Islands (ACP) agreement or the EU Generalized System of Preference program for least developed countries, the Everything but Arms arrangement or EBA. With EU tariffs on apparel imports almost uniformly averaging 12 percent, the benefits of tariff relief are somewhat less than with the US or South Africa.

The ACP and EBA agreements are both governed by rules of origin. In both cases, the EU requires two stages of transformation, so garments must be made of fabric formed in Mozambique and cut and sewn there too. The two agreements differ in provisions for the use of fabric made by other parties to the agreements. In the case of the ACP arrangement, fabrics formed in other ACP countries and cut and sewn in Mozambique before shipment to the EU are eligible for tariff relief. No such cumulating is possible under the EBA arrangement.

The ACP arrangement is set to expire in 2008 and will not be extended. Instead, the EU is promoting Economic Partnership Agreements (EPAs) with its former ACP partners. EPAs are full fledged free trade agreements, so they require reciprocal tariff benefits. The EBA arrangement has no expiration and is promised as long as Mozambique is classified as a least developed country National Strategy of Cotton Textile

Objective to be Achieved

The cotton textile value chain in Mozambique is at an early stage of development. The cotton sub-Sector has suffered from a decade of negative growth, but is now on a gradual upward production trend. The textile and apparel industries are still awaiting a period of new growth. There are several reasons to be optimistic about Mozambique's potential in the textile and apparel industries. First, the extension of the AGOA third party fabric provision through September 2012 provides a window of opportunity to attract new investors looking to take advantage of significant tariff relief in the US market for both textile and apparel products. Second, Mozambique holds out the promise of new labour law reforms and reduced corporate income tax programs for exporters that will bring severance payments and corporate income taxes closer to global competitors. Third, the SSA and the SADC regions are poised to advance to new levels of economic integration as a result of the EU economic partnership agreements (EPAs). With abundant labour and raw cotton, Mozambique stands a good chance of benefiting from greater economic activity in the regional industry.

These advantages alone will not guarantee Mozambique's cotton textile sector a rapid growth path. The government of Mozambique will have to aggressively pursue barriers that remain, such as infrastructure, customs processing, reducing corruption and red tape. Therefore, the targets set for the near term, 2-3 years, are modest. If Mozambique is successful in making major strides improving the business and regulatory environment, these goals should be revised upward.

Cotton Sub—Sector Targets

The improvement of cotton quality and the reduction of contaminants should be the leading priorities. Since the cotton sector is largely under the direction of the Ministry of Agriculture and Rural development, targets should be set in cooperation with that agency. Suggested targets could be to reduce contaminants in seed cotton by 25% over three years.

Organic certification is another target that should be considered in cooperation with the Ministry of Agricultures and Rural Development. A proposed target could be in the order of 5% organic certification in a 3 to 5 year period. Such a program would create significant awareness of Mozambique as a source of high quality organic cotton. Such awareness would also garner the attention of textile and apparel producers.

Textile Sub—Sector Targets

Within a two to three year period, attract at least two apparel manufactures with a record of investing in the vertical integration of textiles and apparel. Such firms would have demonstrated experience operating spinning, knitting or weaving mills as well as garment making-up. Examples of such firms include Nien Sieng of Taiwan (currently producing denim and garments in Lesotho), China Garments (currently producing garments in Lesotho and denim in South Africa), Novel Denim (formally producing garments and denim in South Africa and Madagascar), Gilden Mills a major vertical producer of knit garments. All of these firms have an established strategy of first developing garment infrastructure in a country, using their early experience to evaluate the feasibility and viability of vertical integration into textiles.

The benefit of establishing apparel firms with a proven history of vertical integration is that it greatly enhances the interchange and communication between the firms, local stakeholders and the government regarding alignment of incentives, infrastructure and regulations. These firms will also provide an important vector for learning regarding the improvement of seed and lint cotton quality and grading. If a textile mill were to be established in Mozambique, the benefits would include additional employment of between 200 and 500 workers. On a separate, parallel track, the government should seek out international investors in spinning. The main benefit of seeking out these investors would initially be to improve the interchange between the Mozambique government and investors about the constraints to the establishment of a spinning sector in Mozambique. If a spinning firm was to establish itself in Mozambique, the employment benefits would include 150 to 300 skilled workers. Cotton producers would also benefit from better prices and direct interaction with technical staff within the spinning firms resulting in improved cotton quality.

Apparel Sub—Sector Targets

Within a two to three year period, a cluster of five to ten apparel firms exporting to international markets could be established. The benefits of these firms would include the employment of between 2,000 and 4,000 low skilled Mozambique workers. With proper training programs, supported by the Mozambique government and donors, between 50 and 100 competent line mangers and supervisor could be active within the industry, substituting for foreign staff and providing a major incentive for the otherwise footloose apparel firms to remain in Mozambique.

The establishment of a cluster of apparel firms would be an important first step to the more capital intensive textile industry, since investor will be looking for signals that regulatory and business condition in Mozambique were in a range of economic feasibility. The government of Mozambique would also benefit from the development of this cluster, since it would require close coordination between industry and government providing an important feedback and learning experience to further tune policy to attract investors.

ALIGNING REGULATORY AND BUSINESS ENVIRONMENT WITH COMPETITORS

FRAMEWORK FOR ACHIEVING OBJECTIVES

Mozambique has business elements sought by world-class apparel producers: abundant low-cost labour; port and shipping access; and a stable government and economy. Mozambican producers, however, will have to compete with producers operating in superior business environments. Such environments are characterized by flexible labour laws that support the rights of workers but permit shift work, retrenchment; customs offices that clear imported materials and supplies on a duty-free basis in less than 24 hours; customs offices and ports that clear and load finished products in an environment free of corruption and secure against terrorism; reliable and swift ocean and air transportation networks; access to capital and equipment on a tariff-free basis; freedom from excessive red tape; and a skilled, trained, and productive labour force.

The best way to encourage an export-oriented apparel industry in Mozambique is to put in place policies that promote transparency; reduce the risks and uncertainties of starting a business for local, regional, and foreign investors; and promote productivity, innovation, and workforce incentives.11 Such policies would also encourage development and growth in a broad range of industries and services.

If Mozambique decides to pursue policies to develop textile and/or apparel industries, it will also need to attract foreign investors. This will require ensuring rapid and consistent clearance of goods through ports and customs offices. Apparel producers, who are subject to the rigorous demands of retailing, depend on efficient and delay free supply chains. Indeed, a one-day delay in shipping in the international garment industry is estimated to be equal in cost to a tariff of 0.8 percent (Hummels 1999). Moreover, delays in imported inputs can result in low utilization of labour and machinery and higher average costs as workers are required to wait for essential materials. Because Mozambique is, at best, 30 days by ocean freight (Maputo-Durban-New York) from the U.S. market, it is at a distinct disadvantage compared to East Asian suppliers (12 days) and regional suppliers in the Americas (3-12 days). In other words, Mozambique's location alone can negate its tariff advantage in the U.S. and EU markets. It is therefore essential that hindrances within the control of Mozambique's producers and government—customs, red tape, transportation—be reduced because they can quickly render Mozambique's products non-competitive.

Labour Regulations

Mozambique's inflexible labour laws hinder the development of a competitive textile and apparel industry. The world market for apparel is in constant flux. Buyers frequently cancel orders or place large re-orders when realized commercial sales clear shelves more rapidly than expected. Textile producers also require flexibility. Even under the best of circumstances they must run expensive capital equipment for extended periods, pausing only for maintenance or to switch patterns and designs. Running machinery through multiple shifts allows them to meet surges in demand with higher operating rates rather than risky investment. Higher operating rates and profits, in turn, encourage investment and capacity expansion. Clearly, apparel and textile suppliers require a flexible labour environment, wherein workers are permitted to work multiple shifts (even for a premium) and hiring and retrenchments can take place in accordance with the performance of the industry and supply and demand (most producers retrench labour very reluctantly because the hiring process imposes high costs).

Mozambique has proposed a labour law that will bring some, though not all, of its labour regulations into alignment with international competitors. While the proposed labour laws are a major advancement beyond the current labour laws, especially regarding economic dismissal more needs to be done. The main elements of Mozambique's labour law and the target that should be set to achieve parity with regional and global competitors.

Three items stand out from the four. First, severance for non-economic reasons is much higher than global competitors. With 3 months of notice and 3 months of severance pay for a worker with three years of service, Mozambique's law exceeds competitor's benefits by a factor of six times. The disparity between Mozambique's severance provision and that for competitor's only increases with time, and at yen years, it is far from competitors and an expensive liability for firms to shoulder. Reducing severance to 1 week for every year of service would put Mozambique's labour laws in alignment with competitors.

The second provision that stands out from others is the involvement of the government in the hiring and firing process. As long as the government has the potential to arbitrate in every employment action, producers will be captive to government employees rendering judgment over their production floors. Few producers would approve of this, since they know that it is then difficult to mange their production floors properly. Instead, the government should refrain from intervening in the hiring and firing of workers, unless certain basic worker's rights, such as discrimination, abuse or the right to organize are violated. Finally, limits on paid leave, for any purpose, needs to be set. Without limits on paid leave, factories will incur high absentee rates and resulting higher labour costs. Additionally, with higher absentee rates, factories will be forced to hire even more workers to insure proper coverage. Every worker hired comes with a fixed cost, such as severance and leave liabilities, increasing the overall labour costs for the firm—untenable for a firm looking to operate for a 5- 20 years period. A downward spiral of high absenteeism and rising liabilities is created for the firm.

While the proposed changes to Mozambique's permanent labour laws are a welcome advancement, in their current state, the proposed labour laws do not mach regional and global competitors. In the likelihood that foreign investors will find the labour laws too inflexible and too burdensome, Mozambique will need to consider alternative of solutions, at least for the near term of three to five years. In this regard, several strategies may be taken for immediate action:

- Modifications to the current contract labour laws to allow for more than two to three short term contracts with the same employer;

- An optional "cash-out" provision for producers and workers whereby severance and leave benefits would be liquidated, at least in part, from the start of employment, by paying a portion of severance pay as an annual bonus during holiday periods, at a discount from the current system, but over the entire term of employment. The severance payments would be done at a discount since workers would have cash sooner (the present value is higher) and there would be less uncertainty over any future payout. In essence, the producer would be paying higher wages, but they would not incur severance or leave liabilities beyond regional and global competitors.
- A hybrid of the first two options for designated free zones and exporting firms that provides for greater short term contract flexibility, higher minimum wages and lower severance and leave benefits.

It is important to note that by reducing defined benefit programs such as severance and leave payments, the workers total compensation need not decline, as the augmented wage rate could be defined to proportionately compensate for the reduction in future non-wage benefits. In essence, the national Mozambique labour law would remain intact, but would be paid on cash as you go basis.

Shipping and Customs

The reliability of shipping times is of concern. It can take from 1 to 17 days for goods, imports or exports, to clear customs in Mozambique.13 Goods that arrive late in the major markets are often subject to severe discounts and orders received late can be cancelled.

How important is it to get garments to market on time? Even Asian suppliers, who are closer to the U.S. market than Mozambique, airfreight one-quarter or more of their U.S. shipments.

While there is little that can be done in the short to medium term to increase the frequency of shipping between Durban and Maputo; the fact remains that this is a significant obstacle to the development of a globally competitive industry based on fluid least cost supply chains. This fact has wide ranging impacts on a textile and apparel strategy. What can the government do? First, the government can mitigate the impacts of variable shipping times by ensuring that all procedures within its control are executed with consistency.

Improving the reliability of government administration can go a long way to mitigating the impacts of long shipping times, even though delays due to administration are normally measured in days not weeks.

If producers could plan shipping times within Mozambique, with a high degree of certainty, they could potentially reduce the impacts of shipping schedules.

For example, a producer that knows the departure of the next ship could adjust their production and packing times to coordinate with that ship. If, however, delays in customs cause that ship to be missed, even if only by hours, the whole shipment is subject to the uncertainty of the next departure. The benefits of reliability are, therefore, measured by more than simply the number of hours or days it takes to get administrative tasks accomplished—since the timing of events is critical to the success of a supply chain.

Two recommendations are common across all the interventions:

- Providing customs clearance at the factories, rather than the ports or terminals;
- Electronic documentation.

The provision of customs clearance at the factories is a common approach in many countries with supply chain dependant industries. The approach can be fee based system, where producers can pay for the salaries of the customs inspectors and a fixed amount for their transport (decided in advance on a system wide basis). The important thing is that all costs are paid by the producer, with an incentive to customs personal for responsiveness.

The second proposal, electronic documentation, has a proven track record in major trading countries including Singapore and Malaysia. Electronic documentation is an expensive option and could be implemented gradually over the next 3-5 years. An added benefit of electronic documentation would be to reduce the dependence of Northern customs and port authorities on customs authorities based in Maputo, thereby improving direct shipping times to the North of the country. In the interim, improving communication links between field offices and central offices could help mitigate delays.

If all these recommendations would be implemented, shipping times would be far more reliable, allowing producers an opportunity to adapt to the long shipping times. Once again, this is in contrast to actually reducing shipping times, which would be a longer term goal. While improving the reliability of shipping times would be the major goal of these interventions, shipping times could be reduced by as much as 25% through the implementation of these programs.

Beyond these specific actions, the strategic plan itself should result in competitive shipping times. Major shipping constraints and the manner in which Mozambique's sector strategy can overcome them.

COTTON SUB-SECTOR ACTION PLAN

The cotton sector is formally administered by the Ministry for Agriculture and Rural Development, and this strategy is to compliment efforts by that ministry to advance the cotton sector. In this regard, the cotton textile sector strategy offers several opportunities to increase the incomes of rural households and exports earnings.

Identify Regions and Localities with High Potential

Mozambique is twice the size of California and textile investors visiting Mozambique for a feasibility study will want to quickly identify the localities with the highest potential for vertical integration into the seed and lint cotton sector. Some localities will clearly be unsuitable to their needs because of cotton quality or reliability of the cotton crop. In other instances, the presence of JVs may pose an obstacle. The Ministry of Trade and Industry in partnership with the Ministry for Agricultural and Rural Development should identify localities and key stakeholders with above average seed and lint qualities that are willing to engage with new textile investors.

Agricultural Extension and Learning

One of the key constraints of the seed and lint cotton sectors is low productivity and quality. The cotton textile sector offers a number of opportunities for learning and agricultural extension into the seed and lint cotton sectors. First, introducing new textile investors (spinning or fabric forming) to key cotton stakeholders in Mozambique will set the stage for a dynamic learning process. Cotton producers who understand, first hand, the exact needs of their potential consumers are in an advantageous position to benefit from that knowledge.

The cotton textile strategy also calls for attracting apparel investors with experience with vertical integration. Interchanges between these producers\investors with key cotton stakeholders in the government and private sectors should be facilitated. In this way, all parties will come to a better understanding of the barriers that exist to investment in the cotton textile chain, be they small or large. A process of identifying problems, solutions and innovation will be set in motion.

Improving Cotton Quality and Grading

As long as cotton quality remains low, investors in spinning will be reluctant to take on the challenges of locating in Mozambique. Poor cotton quality will result in poor yarn quality. As long as yarns are of a poor quality, customer rejection rates will be high. The Ministry of Trade and Industry should coordinate with the Ministry for Rural Development on the improvement of cotton quality and grading, by facilitating investor feedback and identifying complimentary programs to improve cotton quality and grading.

Organic Cotton Feasibility Study

A study should be undertaken on the growing market opportunities for organic cotton. The feasibility study should asses market opportunities for yarns, textiles and garments. For the markets with the greatest potential, the assessment should outline certification requirements and the potential for

Mozambique to exploit this market niche. The assessment should address the opportunities and impacts for small farm holders that already use low volumes of pesticides and fertilizers.

Textile sub-Sector Action Plan

The textile sub-sector comprises spinning, knitting, weaving, dying, and finishing—capital-intensive activities requiring imported equipment and materials. A competitive textile sector holds the potential to reduce lead times and production cycles for apparel, lowers materials costs, and enables trade by meeting rule-of-origin required by preferential trade agreements.

Material costs can be lowered through vertically integrated operations, which hold the potential to reduce product cycles and inventories (raw materials and semi-finished products) by aligning production with orders. If local cotton is used, shipping and warehousing costs are minimized. Note that crop seasonality and variations in cotton quality and grades will still require warehousing and imported materials. Thus, national policy should permit cost-effective and flexible sourcing of raw material in contrast to laws that mandate the use of local inputs.

Perhaps the most compelling reason for integrating into textiles is the ability to meet U.S., EU, and SADC rules of origin and thus avoid substantial tariffs on apparel. While the United States provides a derogation on the rule of origin through September 2012, permitting the use of fabrics from anywhere in the world, the EU and South Africa require that apparel benefiting from duty-free treatment be formed in the region. Specifically, garments must be constructed of fabrics formed in ACP or SADC countries to enter the respective markets and be eligible for a tariff reduction. In most cases, the yarns can be from anywhere. After September 2012, apparel exported to the United States must be constructed of yarns and fabrics formed in sub-Saharan Africa or the United States to be eligible for duty-free treatment. In all cases, the production of knit to shape and circular knit garments (such as hosiery) are held to a stricter standard. In the case of AGOA, the fibers of these special knit products must be from the United States or sub-Saharan Africa. In the EU and South Africa, the yarns must be spun in the region in addition to forming fabrics and making up garments.

The benefit of a vertically integrated textile and apparel industry needs to be balanced against the risks, technical challenges, and capital requirements of a textile industry. The textile and apparel industries are now buyer-driven. Investment in textile products needs to be coordinated with end customers; otherwise capacities will be underused and expensive, and unwanted inventories will accrue. Most textile manufactures prefer to limit the degree to which they sell their goods on the open market, since the expense of locating customers can quickly reduce profits and force heavy discounts to secure orders. Therefore, it is of strategic importance that end markets for textile products be secured before an investor will commit to significant investment. In Mozambique, an export-oriented apparel sector is lacking and would be a first step to justify large expenditures for textile development. This is not to say that basic spinning operations could not be established, but they would require coordination with buyers within the SADC region or elsewhere. Foreign investors with established customers and technical knowledge are best suited to these purposes.

Feasibility Study Textiles

While the potential for vertical integration of the cotton textile sector in Mozambique arises from the well known local cultivation of cotton, crucial details of interest to textile investors are largely unknown and are not easily accessible.

A feasibility study should be conducted to locate the most promising locations for the development of cotton spinning. Factors to be addressed in the feasibility study are local provisions for key inputs, including cotton (length, quality and availability, special varieties such as organics), but also addressing:

- local conditions for electricity including the price, quality and consistency;
- Access to clean and consistent water supplies;
- The availability of waste water treatment;
- Access to ports and transportation costs.

The feasibility study should not restrict itself to retired textile plants, but should be open to green field investments in new locations, should conditions permit. The feasibility study should identify the top 3 locations in the country for textile investment, based on cost and infrastructure criteria.

INVESTMENT PROMOTION FOR TEXTILES

An active investment promotion program is needed to facilitate the flow of knowledge about Mozambique as a location for potential investors and to provide feedback to local producers and officials about constraints to investment in the sector.

The first step in the investment promotion program would be to package information from the feasibility studies into a brief of brochure that would be marketed to potential investors. Lists of key contacts and facilitators would also be included in the export promotion brochure. The brochure should be made available through the CPI and MIC, but one central contact should be designated in Maputo for facilitating the flow of information to potential investors.

Funding permitted, an active marketing campaign would be conducted, targeting key investors in the region and globally. A textile marketing firm will have to be hired for this purpose. At first, it would be best to target investors in the spinning sector, since the capital requirements are lower and shipping times are less important than with finished fabrics.

The marketing campaign would have to fund trips to countries with key investors, such as South Africa, Pakistan, India, Europe and Brazil. A follow up plan would have to be in place to facilitate visits by these investors, should they find the opportunities worth following though on. In all cases, a feedback mechanism should be in place to gather investor's impression, concerns and interests.

Targeting Producers with Vertical Integration Experience

In coordination with the apparel sector investment promotion program, specific out reach should be made to apparel investors with experience in vertical integration. Several such producers exist in the region such as Nien Sieng of Taiwan in Lesotho, China Garments in South Africa and Lesotho, and Novel Denim formally located in South Africa and Madagascar.

With the new AGOA provision for wholly originating textile products, producers of these products with experience in vertical integration should also be pursued. Such investors could be from South Africa (Frame Textiles being the largest), the US (West Point Stevens, Burlington Mills, Cone Mills etc.14), Europe and Asia.

Developing Niche Products

Several areas of nice product development should be pursued including:

- Organic cotton products.
- Products benefiting from Mozambique's longer staple fiber length, such as those using high count yarns, for example high quality shirting and sweaters;
- Products of African design;

The development of socially conscious materials continues to grow exponentially and Mozambique could find a niche in this area. Identifying buyers and producers looking to fill these niches will be challenging, but the development of organic cottons and the publicizing of Mozambique as a viable location for souring these products would be critical.

Apparel Sub-Sector Action Plan

Attracting foreign investors for development of an export-led apparel industry could help boost job creation and reduce poverty in the short (1-3 years) and medium (3-5 years) term. Developing the industry to serve domestic markets is not likely to succeed at present. The local consuming population is small (18 million) and poor, able to afford only cheap imports of used apparel. Without such imports, Mozambique's poorest would have to divert scarce income from other necessities such as food, water, education, and medicine. The benefit of a few new jobs in an apparel industry dedicated to domestic markets would not justify the negative effects on cost of living. An export-led strategy recognizes that foreign markets offer far greater opportunities because of their size and wealth. To support production for the local market, the Government of Mozambique can contract for uniforms and provide assistance, including marketing assistance, for production for local niche markets, such as for Capulanas.

Because of the poor state of local transportation and infrastructure, and the importance of both to foreign investors considering locating in Mozambique, export-oriented industry will have to be near reliable infrastructure and have good access to ports and roads. As local infrastructure improves over the medium to long term, geographic diversification may be achieved.

While the short- and medium-term goal should be development of a competitive apparel industry, integration into textiles can be aided by supporting apparel producers who have experience integrating into the textile production chain.

Investment Promotion for Apparel

With the AGOA third party fabric benefits recently extended through September 2012, Mozambique has a renewed opportunity to seek out investors in the apparel sector. As the EU EPA negotiations continue to unfold, new opportunities for exports to that market will also develop. The development of the apparel sub-Sector is also a key link in the development of the textile and seed cotton sub-Sectors and its development should be a top priority. It's also important to bear in mind that as Mozambique makes progress on key aspect of the business enabling environment, opportunities with investors will continue to develop, but the time to start seeking investors in the apparel sector is today. As a first step in an investment promotion program, Mozambique will need to develop a promotion package, similar to those for the textile sector, but more aggressive.

Several factory shells in the Beleluane free trade zone should be established for marketing to potential apparel investors. The shells would ideally be paid for and rented at a steep discount for the first five years for the first firms that set up there. In the absence of resources for this purpose, several established factory sites outside the free zone15 would be identified for lease at subsidized rates.

Realizing investments will require more than a package of laws, regulations, and incentives suitable to apparel investors. Few foreign investors are familiar with Mozambique as a location for producing apparel. Mozambique will have to promote itself and its advantages and encourage a flow of information between government and potential investors. The Government of Mozambique needs to create an investment promotion package that prospective investors can respond to by requesting adjustments that address shortcomings in Mozambique, or by expressing further requirements to be addressed by government officials. The government can then refine the promotion package.

A governmental or quasi-governmental organization (such as the CPI) can assemble an initial investment promotion package and contact foreign investor prospects such as South African producers, Asian transnational producers (such as those already in the region), and producers of vertically integrated knit apparel in the United States, Europe, and South Africa. The organization should also contact producers with direct ties to buyers to minimize the risk inherent in contract production, although this may be unavoidable in many cases. The initial package needs to take into account the considerable risks and uncertainty of being first investors in this sector, which has few examples of success. In this regard, the objective should be to create an initial cluster of 3-5 apparel firms. The success of these firms over the short and medium term will signal to other investors that Mozambique is ready to meet their needs to be competitive in the global market. This initial package, offered to the first firms to locate in Mozambique, should include extended income tax holidays and preferred buildings and infrastructure.

Further requirements may emerge, but the entity responsible for foreign investment promotion (presumably CPI) should be required to work closely with foreign investors to identify their interests and respond to their concerns till deals are struck. Financial incentives, however, are rarely the honey that attracts investors in this industry and should be avoided. Stable investors will be attracted by the likelihood of long-term profitability and security. Activities to attract these investors should focus on non-monetary measures within the control of government authority.

After the initial cluster of firms is attracted, the focus should shift to creating an association that will continue addressing the concerns of foreign investors and producers. The government or private investors can expand facilities within the free zones, while charging rents commensurate with costs and maintenance for buildings and infrastructure within the industrial park(s). The central government would secure and improve transportation, ports, and communication infrastructure.

PRODUCTS GROUPS

There are three distinct apparel product groups: Fashion products, fashion-basic products, and basic products.16 Within this group of products, certain types of products have a history of vertical production and employing low skilled low wage labour. The following sections review each of these products and their suitability for production in Mozambique.

Fashion Products

The marketing costs and long lead times of fashion items do not serve Mozambique's employment objectives. While high fashion holds long-term potential, it requires considerable marketing and expense. In the high-fashion market, buyers favor established suppliers with a solid history of networking and contacts. Networks are developed with established expatriates in market centres such as New York and Los Angeles. Mozambique's near-term potential in this segment is limited.

Fashion-Basic Products

In this segment delivery, services, quality, and product range are more important than cost. Products and producers include most name brand apparel producers, such as Liz Claiborne, Levi Straus, and major department stores. Service has to be balanced with cost in this category. The supplier's ability to manage high-level services while controlling costs is a primary requirement. Buyers usually seek "full package" producers that can manage every detail of the supply chain, from fiber to the showroom floor. These services include design, quality control, cutting, sewing, sourcing raw materials, logistics, and labeling. Local industry and infrastructure must be able to support such services as well as shorter than average lead times for re-orders. Apparel companies successful in this segment are more closely coordinating with textile producers and retailers and often nominate textile purchases from a select group of producers—limiting the potential for textile integration in Mozambique.

Because of wide product ranges, vertical integration can be as limiting as it is beneficial. Given its distance from the U.S. and EU markets, and its present lack of a skilled and experienced service infrastructure, Mozambique would have considerable difficulty attracting direct suppliers for this market segment. However, contract sewing firms (known as CMT firms) managed by investors in Asia, close to the main supply and logistics chains, will locate limited production in locations such as Mozambique or Lesotho., if costs are competitive and tariff advantages are compelling.

Basic Products

To keep costs low and predictable, competitiveness in basic products requires political stability, a good business environment, and sound investment laws. While producers are willing to trade market proximity and lead times for lower average costs, they still require a predictable and reliable shipping schedule. Production inputs and exports should rapidly clear customs, and customs procedures should not tie-up working capital in complex duty rebate programs. Government corruption and the "unofficial" costs of getting a product to market should be minimal, but above all, predictable and stable. Obscure government processes and regulations and complex labour laws make it difficult to plan production costs and schedule production. Producers do not require significant local services for design pattern-making and pre-production process because apparel lines are standardized and pre-production is often coordinated from central offices. Production volumes are high, though, usually requiring 2,000 or more machinists to service the U.S. market, and somewhat less for the EU and South African markets. The size of efficient establishments has been increasing; a minimum establishment size of 5,000 or more to service the U.S. market is not unusual.

Mozambique can achieve rapid growth in the basic products segment—but it can also experience decline just as rapidly. Competition is intense. The major apparel manufacturing countries, from China to India to Mexico, are all trying to build market-share in basic products. The basic product segment is the largest product market and encompasses every product, from trousers, hosiery, undergarments, dress, shirts, skirts, and suits.

Attracting producers and product categories that have a history of backward integration into textile products and processes could spur development of these industries and fully leverage the natural advantages of Mozambique's seed and lint cotton industries.

PRODUCTS FOR VERTICAL INTEGRATION

A key marketing point and potential opportunity for an investor to locate in Mozambique will be the potential to integrate production into textiles making use of Mozambique's excellent cotton fiber length and contributing to a reduction in potential lead times and costs.

Special attention, therefore, should be given to attracting apparel producers with a proven record of vertical integration in cotton textiles including weaving knitting or yarn spinning. Examples of producers with this experience investing within the SSA region include Nien Sieng Denim with garment making and denim weaving investments in Lesotho. China garments with denim weaving capabilities in South Africa and garment manufacturing in Lesotho.

While both of these investors have chosen to produce heavy weight cotton pants, Mozambique should also consider products that require lower than average capital investments and target producers of these products in the investment promotion programs.

- *Cotton Knit Undergarments.* The making-up of knit undergarments is associated with low profit margins because value added per garment is low. Major producers of knit undergarments often seek sites for long-term vertical integration, which allows them to capture more value added while controlling processes and costs. Producers often "test" production locations with cut-make-trim (CMT) operations with medium and long-term goals of establishing vertical or contract spinning and knitting operations. Capital requirements are lower than those of weaving operations, but are still substantial, at between US$ 30-40 million for a plant with new equipment. Producers include Hanes, Fruit of the Loom, Gilden Mills, and Russell.

- *Cotton Hosiery.* U.S. imports of cotton hosiery exceeded US$ 600 million in 2005. Major exporters to the U.S. market include suppliers in Korea, Mexico, and Central America.
- *Cotton Knit Shirts and Sweaters.* U.S. imports of knit shirts and blouses exceeded US$ 2 billion in 2004. This segment includes a wide variety of vertical and non-vertical producers making everything from basic products to high-fashion items. Vertical integration into knitting and spinning can provide distinct cost and quality control and lead time advantages.

Products Suited to Low-Skill Labour

Knit products offer good possibilities for vertical integration based on relatively low capital outlays but do not make full use of one of Mozambique's competitive advantages: low-skill labour. Low-skill labour constitutes 25 percent or less of the f.o.b value of these products. For cotton hosiery, the value added by labour can be 10 percent or less. Products that use more labour *and* more skill include dress shirts, suit jackets, and women's ensembles. Low-skill labour is more competitive when employed making heavyweight cotton trousers and denim jeans. Heavyweight trousers often require 30-50 percent value added by labour because of the large number of components (pockets, liners, waste bands). Since much of the sewing is done inside the trousers and fabrics are usually a solid colour, details are less critical and require less skill. Heavyweight trousers are often standardized (although fashion denim jeans are a major exception) with only small changes in styling from year to year, if ever. Buyers of basic heavyweight trousers include most major low-priced retailers, including K-mart, Wal-Mart, and Target as well as name brands such as the Gap, Levi, Lee, and Wrangler. Attracting producers of these products would be easier, since they leverage Mozambique's assets in abundant low cost, low skill labour.

Products for the South Africa Market

Making up a wide range of apparel for the South Africa market also holds potential for Mozambique. South African labour costs are much higher than Mozambique's. Lesotho is currently experiencing a surge in investment from South African producers.

Used Clothing

Countries such as Kenya, Nigeria, Ethiopia, and South Africa have banned used clothing imports with an eye to developing local textile and apparel industries. For this strategy to succeed the protected industries must be serving a large number of consumers with modest to high incomes. Even where the income threshold can be me met for a small segment of the population, challenges still exist from the lack of scale and specialization. For example, South Africa's producers try to be everything to everybody, which causes inefficiencies, higher prices, and lower quality. South African producers have also been under intense competition from imports despite tariffs of more than 40 percent on apparel. Nigeria does not have an affluent population, so the import ban is borne by the poorest that can barely afford proper nutrition, clean water, education, and medicine, much less high-priced, low-quality, locally produced clothing. Who benefits? The answer is a few local producers who employ only a small fraction of the total population and smugglers who take their cues from higher prices and the desperate needs of the poor. If implemented in Mozambique, an import ban would erode one of its natural competitive advantages: a low cost of living that makes low incomes tolerable. In addition, any increase in employment from the growth of a local apparel industry must be weighed against the loss of jobs in imported used clothing. Thousands of workers sort, grade, clean, and organize the used clothing. Banning used clothing is unlikely to reinvigorate enough local suppliers to justify the negative effects on employment and poverty and would not lead to a competitive industry. Mozambique is a poor country, with more than two-thirds of the population below the poverty line; and HIV\AIDS has skewed the population to the very old and the very young—45 percent of the population is 14 years old or less. A population of this demographic is addressing very difficult issues in fulfilling basic human needs at the lowest cost possible. Given the minimal benefits of banning used clothing and the negative effects on poverty, industrial policy should eschew a ban except in relation to certain niche products (*e.g.*, uniforms, capalanas, and certain artisan products).

Small and Medium Enterprises

Small and medium enterprises are best suited to serve as supplier or contract firms to larger export oriented firms that can manage quality control and customer service. Small and even micro enterprise firms can engage in the support of larger apparel firms in three major ways:

- As subcontractors for the manufacture of apparel;
- As suppliers of accessories, trims and packaging materials, such as boxes, hangers, and bags;
- Service contractors for sanitation, security and administrative task.

The development of small and medium enterprises for these purposes is facilitated by the clustering of apparel firms in a particular area, since servicing firms in a diverse geographic market is more difficult for small and medium sized firms. A strategy that encourages a clustering of firms, therefore, has an added benefit in for the development of small and medium sized apparel firms.

Small and medium sized firms may also serve the local market for tailoring and the manufacture of customized garments for the local markets.

FREE TRADE ZONE AND CLUSTER DEVELOPMENT

Mozambique's laws governing free trade zones are relatively well developed. They allow any firm that exports more than 85 percent of its production and that have 500 or more employees to take advantage of duty-free inputs and the special income tax incentives. Mozambique's free laws also provide for the creation of free trade zones, with lower employee requirements in a firm, but only one free trade zone exists in Mozmabique—Beleluen. The current approach favors decentralization and is in contrast to many countries that provide designated industrial zones where all companies within the zone are companies that export and receive special benefits. Creating special, centralized, industrial free zone, in the South of the country, can benefit Mozambique by providing:

- Access to the best shipping routes in Mozambique, including ports, roads and international airports;
- An established free trade zone authority for monitoring shipment movements and customs procedures and to provide valuable feedback to the government on performance;
- Infrastructure and the development of new factory shells;
- The potential to develop special rules regarding taxes and labour laws for a well defined geographic location;
- A focal point for eliminating red tape and slow customs procedures;
- A focal point for marketing efforts providing easy feasibility analysis for potential investors;
- The potential to link with small and medium enterprises since employment requirements are far lower than for firms outside free zones.

Centralized free zones could also facilitate coordination with South African customs officials and address security and customs concerns in other countries. Today, South African customs officials are very suspicious of shipments from Mozambique, no matter what the shipping documentation or destination, including onward shipment to other countries. Having central export processing zones, with a high level of integrity and bonding of goods for re-export through South African ports, could reduce shipping times.

There is now however a lack of factory shells in the Beleluen free trade zone which needs to be addressed. The government of Mozambique should consider investing in three factory shells ready to house apparel firms of 300-500 workers. These shells would have to be leased to new investors at subsidized rates. The government of Mozambique could also pursue other tax and regulatory advantages especially for this free trade zone, including elimination of income taxes for the fist ten years and potential, more flexible labour arrangement.

INTERNAL CONSTRAINTS AND ELIMINATION OF TEXTILE

The elimination of textile and apparel quotas in 2005 have given global investors unfettered flexibility when choosing where to locate and invest. With literally dozens of countries seeking to lure these investors to boost their industrial production and employment, Mozambique must consider carefully its internal constraints to attracting textile and apparel sector investors. Many countries can count liberal US and EU market access as a benefit, therefore, liberal market access can not compensate for excessively restrictive local regulations and poor infrastructure.

The Mozambique's internal constraints in relation to other countries regarding three key factors to textile and apparel industry investors:

- Getting goods to market reliably;
- Flexibility of labour laws regulations;
- Taxes and incentives for the sectors
- Export processing zones.

Bringing these key factors into closer to alignment with major regional and global competitors will give a significant boost to the establishment of both textiles of apparel industries.

The following sections review the major regulatory and business environment constraints in Mozambique as they pertain to the cotton textile and apparel value chain. Several constraints in the regulatory and business environment of Mozambique are common across sectors and are reviewed first. The sections on the business and regulatory environment are followed by a review of sector specific restraints.

BUSINESS AND REGULATORY ENVIRONMENT

Labour Laws

Labour costs are a significant component of manufacturing costs, perhaps only exceeded by the cost of raw materials (cotton), yarns and fabrics. At $50 per month the minimum wage in Mozambique is low when compared to regional and global competitors. However, wage rates are the starting point for an analysis of total labour costs, which include many non-wage costs such as vacation pay, sick leave, overtime, and severance pay. Moreover, there are many regulatory issues such as registering workers, the difficulty of retrenching them, and the government's involvement in the hiring and firing process. How these non-wage costs balance with wages can be a key factor determining where international investor in textiles and apparel is likely to locate3. Therefore, it is important that Mozambique's labour laws compare with those of major producers of textiles and apparel.

Labour laws and regulations can be organized into five major categories: severance and notice; vacation and sick leave; government involvement; and apprentice laws. The remainder of the columns indicating Mozambique's current law and the laws of two competitors, Lesotho and Cambodia.

The first category, severance and notice, often referred to as the "flexibility" of the labour law, are the conditions under which a producer can retrench workers. Two categories of severance are defined. First economic dismissal addresses the needs of producers to reduce production and the work force because of means outside of their control, such as an economic recession or a reduction in orders. The second type of dismissal address the occasion when producers wish to retrench workers for any reason other than those covered under the economic dismal category, such as poor worker performance. The case for non-economic dismissals. In Mozambique, a producer is required to notify a worker three months before they are retrenched *and* the producer is still required to pay the worker three months severance pay. For all intents and purposes, notification and severance are equivalent, since few producers would maintain workers on the factory floor who have been given notice. This means that a producer in Mozambique that wishes to retrench a worker for non-economic reasons must pay 6 months in wages to the dismissed worker. In contrast, the same producer located in Lesotho must pay a total of 2.5 months severance and notice and a producer in Cambodia must pay one month severance and notice for a worker with the same seniority. In other terms, a producer in Mozambique is required to pay six times the severance cost in Mozambique compared to Cambodia.

The government of Mozambique is currently considering revisions to the labour laws to address some of these concerns. The total notice and severance that a producer must give a worker under the current labour law and under the proposed labour law. Each bar is associated with the seniority of a worker-from six months to ten years on the job. The first bar indicates the current labour law, and the second bar (blue) illustrates the new labour law. In the period between six months and three years, the new labour law appears to be competitive within the textile and apparel industries. At year three, the current and new laws are nearly equal. After ten years, the new labour law requires greater severance payments by the producer than the current law.

The proposed labour law is an improvement over the current law, especially in the short to near term. After three years on the job (not uncommon for an apparel firm) the laws are nearly identical, and the same cycle of disincentives and low productivity are set in motion4. Some counties, such as South Africa and Indonesia, maintain similar laws, but their impracticality is reveled by the proliferation of subcontracting regimes whereby producers locate to remote areas where enforcement of the laws is lax, or loop holes in local laws, or free trade zones (FTZs) laws are created to relieve the economic pressure of these regulation. In some countries, such as South Africa, elaborate systems of hiring and firing workers on a cyclical basis is conducted to ensure the workers do not accrue severance benefits and that the textile and apparel factories do not accrue large severance liabilities.

Today, there is uncertainty as to when Mozambique's proposed labour law might be approved. It remains to be seen if Mozambique has actually done enough to attract investors. This analysis indicates that the proposed labour law has not gone far enough, especially in regards to severance for non-economic reasons, with reforms to put it on par with its major competitors in the textile and apparel sectors.

The second item is the level of government involvement in labour disputes that do not violate the ILO basic code of workers rights. In most countries with growing apparel industries, the government does not get involved when an employer dismisses a worker, as long as the proper procedures of the law involving severance and notice are lawfully applied. In Mozambique, the employer is required to notify the government of its intention to dismiss a worker, and the government intervenes or arbitrates. Producers in Mozambique find this process extremely cumbersome, and any fault in carrying out this process favors the worker. Once again, a process is put in place that raises the cost to the producer of dismissing workers not performing to the level of their peers, since the worker knows the process will work in their favor. Productivity and management of the factory, therefore, suffer, in addition to the producer having to disengage from their primary concern—running a factory and meeting the buyer's needs.

Paid vacation and sick pay also raise the costs of employment. A worker in Mozambique is entitled to six weeks of paid vacation (1.5 months) after three years of service with an employer. The standard for vacation in countries with strong and vibrant textile and apparel industries is one day of paid vacation for every month of service or a little over two weeks of vacation (0.5 months) for the same worker. Therefore, a producer in Mozambique must pay three times the annual leave costs as compared to the regional and global standards presented here. Likewise, paid sick leave is allocated at the same rate as vacation, one day per month of service. In contrast, the Mozambique labour law appears to set no limits on how long a factory must pay a worker on sick leave, assuming reasonable justification is established.

Finally, regional and global apparel producing countries have provisions for apprenticeship programs and new workers to a factory. Apprenticeship programs allow producers to pay a reduced rate for workers and allow management to dismiss an employee without cause in the first few months of service. This is important to a factory, since a worker's skill and aptitude for a job can only be evaluated as they are observed on the production floor. Workers that do not meet the standards for the job are dismissed.

MARKET RELIABLY AND EFFICIENTLY

SHIPPING AND DELIVERY TIMES

Getting goods to market reliably and efficiently is critical to the establishment and growth of textile and apparel industries. The value of apparel and textiles is tied to fashion trends and seasonal sales patterns. Textile and apparel products, therefore, have more in common with perishable produce, such as fruits and vegetables then they do with metals or machinery. Cotton producers that receive fertilizers and pesticides late produce less cotton. Spinners, weavers and knitters that do not have access to spare parts, dyes or technical samples of yarns and fabrics loose orders. Apparel producers that deliver products late have their payments reduced by 20 -50% and key customers are lost. Long and variable shipping times in Mozambique will limit access to materials and supplies for the various links in the supply chain, and this constraint will be an important determinant of the types of investments that will be made. While Mozambique's cotton sector offers potential to reduce dependence on established international supply chains and reduce shipping and lead times, it must be recognized that many challenges still exist for shipping within Mozambique.

International Shipping

The value of apparel and textiles is tied to fashion trends and seasonal sales patterns. Goods that arrive late are usually worth a fraction of goods delivered on time. The maximum value of a textile or garment is only fully realized when it is delivered to the end user or the retail floor when it is needed, not one minute later. Goods that do not meet the timely demands of consumers are often heavily discounted. Worse yet, an empty shelf or an idle loom or knitting machine represents a lost sale, higher overhead for the producers and lower profit margins. Even for standard garments, such as white t-shirts, rapid and fluid transportation is required; if less for the fashion appeal, then for the management of material and inventory costs such as lint cotton, yarns and dyes in an extremely competitive segment of the market. For these reasons, reliable and rapid shipping times are crucial to the textile and apparel sectors.

The time required (in weeks) to manufacture and ship a garment starting from the time of order placement with an apparel firm in Mozambique. The same data are shown for Lesotho, a close regional competitor with a successful making-up industry. The fabric and materials are sourced from China; a common practice in sub-Saharan Africa (SSA), since AGOA permits the use of non-regional fabrics. The distinct processes are along with summary data:

- Manufacture of fabric from order (in China);
- Shipping of fabric from China to South Africa;
- Unlading of cargo and customs inspection;

- Making up the garment ;
- Pre-shipment inspection;
- Delivery to South African port.

Three data points are presented for each process for each country—the minimum time required, the maximum time required, and the difference between the maximum and minimum or variance in shipping time. The variance figure reflects the reliability of shipping. The minimum time from order to delivery of the garment to the dock of the closest South African port is 11.6 weeks for Mozambique and 10.8 weeks for a producer in Lesotho. These shipping times can be contrasted with shipping times from Asian apparel firms, which can deliver a garment to a US port in 10-12 weeks, on average. It takes on average 3 weeks to ship a garment from South Africa to the US, therefore, both Lesotho and Mozambique are one and a half to two and a half weeks behind Asian competitors in delivery time—if imported fabrics and materials are employed.

Tariff free access for the producer in Africa largely explains why sourcing agents for the US would incur time and shipping costs beyond those found in Asia. More important than the minimum delivery time is the reliability of shipments or the variance, the difference between the minimum and maximum shipping times. All manufactures and shippers experience delays at one time or another; these can be due to seasonal volumes, unexpected events or natural causes. At other times, it can be due to routine government or port activities, inconsistently applied such as processing of shipments through customs offices and variable vessel departures.

The maximum shipping time and hence the variance data delays due to these routine activities. In the case of Mozambique, a producer can guarantee a shipment only within a 3.7 week window, nearly one month. In contrast, a Lesotho based producer can routinely guarantee a shipment's delivery within half a week. Where do these variable delays come from when producing in Mozambique?The vast majority, or 3.4 of the 3.7 weeks, comes from shipping delays from the Port of Durban.

By far, the longest delays come from the Durban to Maputo shipping schedule, which is really not a schedule at all, since scheduled container ships only load cargo for Maputo if there is sufficient volume to justify stopping in Maputo port and are sufficient to cover warfage fees.

This means that a container arriving from Asia must wait in Durban for an indefinite time before being loaded on the first available ship to Maputo. Producers and freight forwarders estimate this time between shipments from fourteen to twenty-one days on average. The alternative for a producer, waiting for time sensitive cargo, is to pay eight times more than ocean freight cost, and ship the container by truck. Not only is this a more expensive option, it is a riskier one, since cargo trucks are subject to hijacking and theft. The producer is presented with a Hobson's choice; receive materials late by ship, or pay more, and possibly receive them on time, or not at all by truck.

While there is little that can be done in the short to medium term to increase the frequency of shipping between Durban and Maputo; the fact remains that this is a significant constraint to the development of a globally competitive industry based on least cost supply chains. This constraint has wide sweeping impacts on any national or firm based textile and apparel strategy.

International air cargo can be shipped from the Maputo airport, but air freight rates from Maputo are reported to be much higher than routes of similar distances outside of Mozambique due to a lack of competition between air freight carriers at the Maputo airport. Air freight, therefore, should not be considered as a principal option for shipping goods internationally. South African ports and air terminal may be accessed via a good road system in the South, but are practically speaking, inaccessible from the North of the country.

Local Shipping

Mozambique is a country that is approximately 2,000 kilometers long and nearly twice the size of California. East-West transport linkages for rail and trucking are well developed in the areas of Beira, Maputo and Nampula. However, North South linkages are lacking. No rail lines run the North—South meridian of the country. Roads are of variable quality and bridges are missing from major river crossings. A report by Global Development Solutions in 2005 reported that a truck originating in Nampula province, known for its cotton production, would have to navigate three countries just to arrive in Maputo. Ocean shipping from Northern Mozambique to Southern Mozambique can cost as much as shipping from Asia to South Africa. However, local shipping lead times hold the potential to be considerably better with no customs clearance required.

TEXTILES AND APPAREL INDUSTRIES

CORPORATE TAXES AND INCENTIVES

The textiles and apparel industries are intensive in their use of unskilled labour and many developing countries seek to reduce unemployment by encouraging these industries. Moreover, the textile and apparel industries often offer the first opportunity for many countries to kick start their industrial development programs, since process and supply chain dependant industries are a proving ground for the development of more highly skilled labour. This has led to many countries to reduce, and in most cases, eliminate taxes on corporate profits to encourage foreign investors in these strategic sectors. The premise of these policies is that reducing tax rates on companies is a cost effective means for generating low skill jobs, reducing poverty and developing a skilled workforce.

It has become the *de facto* standard in the textile and apparel sectors to offer complete relief from corporate taxes for a period of time that ranges between three to five years, with even greater benefits tied to the importation of new technology and expatriate expertise. In all cases, countries offer foreign investors complete relief from corporate taxes at least for some period of time. In contrast, Mozambique offers a 60 percent reduction in the standard corporate tax rate of 32 percent, or a 12.8 percent tax on corporate profits. Mozambique's tax on corporate profits is, therefore, uncompetitive.

Some countries find that the elimination of corporate taxes does not provide sufficient incentive or they project that when these benefits expire, there will be little incentive for the companies to stay. In some countries, wider incentives packages are offered to address these concerns. These incentives vary considerably across countries.

There are, however, common elements to many of these programs. These elements include finance for worker training programs that will enhance the skills of workers and the productivity of the companies for the medium to longer term. In some instances, the government provides basic training for operators (Cambodia, Dominican Republic, Egypt, etc) and in others, the government shares in the companies training costs (South Africa, Lesotho, Malaysia). In the case of Mozambique, there is little in the way of systematic support for corporate or industry training initiatives.

Although no formal survey of training programs has been carried out in the textile or apparel industries, producers are quick to note that training programs implemented in the factories are more effective than government run training institutions, since government run institutes rarely keep up with the pace of changes in industry. Therefore, initial evidence indicates that programs that co-finance factory implemented training should be favored over government implemented training programs.

Cotton sub-Sector Constraints

The raw cotton sector forms the base level of the cotton textile manufacturing supply chain. Inflexible labour laws, corporate tax relief and international shipping times are less of a concern for the 350,000 small farm holding than in the downstream manufacturing segments of the supply chain. Nevertheless, institutional constraints specific to the seed and lint cotton sector have limited the development of the cotton textile supply chain. The constraints on the seed and lint cotton deserves special attention, since experts agree, the potential for growing high quality longer staple cotton in Mozambique is significant.

Current supply chain limitations in the manufacturing segments, such as long lead times and high shipping costs could be overcome with an efficient local cotton and textile supply chain.

Low Seed Cotton Yields

Seed cotton yields in Mozambique rank among the lowest in Africa and are between one-third and one-half of world standards. Factors constraining the seed cotton yields in Mozambique include:

- Underutilization of fertilizers and pesticides;
- Low yielding cotton varieties;
- Size of land holdings (small);
- Lack of agricultural extension programs; and
- Lack of finance.

COTTON CONTAMINATION AND PROPER GRADING

When small farm holders harvest cotton, numerous opportunities arise for non-fiber contaminants to enter the raw fiber. Once these contaminants enter the raw fiber stock, they are difficult to remove and result in yarns and fabrics that are of low quality. Common contaminants include artificial fibers from bags used to harvest seed cotton and non-cotton fibers and dirt. The problem of contamination is compounded when uncontaminated fiber from one farm is mixed with contaminated fiber from another. Seed cotton from Mozambique is known to be among the most contaminated in Africa.

Cotton fibers come in various qualities, so proper grading and sorting of cotton before ginning can greatly increase the price paid to farm holders, increase yields of ginned cotton while reducing waste. In Mozambique, cotton grading is inconsistent and unreliable. Grading of cotton is often done far from the fields, and only after numerous farms have consolidated their harvests, eliminating the possibility of sorting out the best cotton from the worst.

Concession System

Three joint ventures (JVs) between the government of Mozambique and private manufactures accounts for more than 50 percent of the seed and lint cotton sector in Mozambique. The remainder of production, not under the control of the JVs is controlled by less then a dozen private concessions. The concession holders manage geographically defined markets. The concessions are responsible for all agricultural extension, processing, finance and marketing of seed and lint cotton. As such, they operate as virtual geographically defined monopolies. Cotton farmers must sell their cotton to the concession company controlling the area they produce in.

Prices for cotton lint are set by the Comissao Nacional de Salarrios e Precos. Prices are set to insure a viable cotton crop (minimum pricing) and to stabilize the income of farmers. The price setting process rarely acknowledges variations in cotton quality between farms or even regions. An important incentive to improve cotton quality is therefore eliminated. While the concession system has been credited with the revival of the cotton sector in Mozambique, the concession system has reached its limits. The system has been blamed for the low yields and the failure to manage cotton quality and improve outreach to small farm holdings. Moreover, the lack of competition has eliminated most incentives by small land holders to invest in their own crops and farms, since prices are uniformly set. It is inescapable that the reportedly low price of Mozambique's raw cotton is linked not to efficient low cost production, but the institutions that mange the prices and activities or producers.

Transport Linkages

The majority of Mozambique's seed and lint cotton is produced in the North of the country. Importantly, the best infrastructure for the development of manufacturing industry, certainly apparel firms, is in the South of the country. On the transportation constraints, these two regions are separated by approximately 2,000 km of poor transportation infrastructure. This limits value chain linkages within the political boundaries of Mozambique.

Textile sub-Sector Constraints

The textile sector is characterized by mechanization and high capital intensity as compared to the cotton lint and garment sectors. In order to insure the highest possible return for expensive machinery, efficient textile factories must run 24 hours a day, 7 days a week for 355 days a year. Factors limiting factories abilities to meet competitive levels of high capacity utilization, such as labour laws limiting overtime and inconsistent electrical supplies are constraints to the development of the sector. Today's textile sectors, including spinning fabric forming and finishing are also dependant on a high degree of coordination with end users to assure quality specifications are met.

Labour Laws

Mozambique's labour laws generally do not restrict individual worker hours beyond a 48 hour work week. It is possible for a factory to operate 24 hours a day for seven days a week without violating labour laws. Multiple shifts would, of course, be required.

Mandatory holiday, leave requirements and severance laws could limit investment in the textile sector. While apparel firms generally have seasonal fluctuation in orders that can be coordinated to moderate the impact of leave requirements, textile firms will have to hire more workers to operate factories continuously while conforming to Mozambique's holiday and leave requirements. It is difficult to quantify the impact of these laws on the cost of textile firms, since certain components of the labour law regarding sick leave are indeterminate. Producers in Mozambique currently estimate absenteeism of workers between 7 and 12 percent as compared with international standards of about 2 to 3 percent.

Severance laws also pose a challenge to textile firms, since the average life of a textile firm is between 6-10 years. Under the current labour laws regarding economic dismissal of workers, a textile firm would be confronted with rapidly growing severance liabilities as it was entering a sensitive period of profitability and potential review of re-investment in new machinery. In essence, a textile firm would have to build these severance payments into its feasibility forecast over the life of the project (6-10 years). Importantly, Mozambique's proposed labour laws reduce the cost of economic dismal of workers by more than 50% providing a more conducive environment for investment and future re-investment.

Utilities (Electricity and Water)

Modern textile plants are heavily automated with spinning, weaving and finishing equipment putting relatively high loads onto the electrical grid. Electrical costs in Mozambique of US$.03-US$.04 are approximately half that of major competitors in India and China. This stands as a potential benefit to textile manufactures, but only if the electrical supply is consistent and of a relatively high quality. Mozambique's electrical power, while inexpensive, is not of a regular and consistent quality. Mozambique's electrical grid is known to experience frequent outages, dips and spikes in voltage that can damage modern textile equipment. Power outages not only reduce capacity utilization but they can impact the quality of yarns spun, fabrics formed and finished. The advantages of low cost electricity can be quickly erased by the need to maintain backup power generators.

Moreover, the quality and cost of power can vary greatly from the South to the North of the country. The South of the country has a more developed power grid (serving the aluminum smelter MOZAL) than the North, where the majority of cotton is harvested.

A textile industries requirement for clean water derives from the need to bleach and finish yarns and fabrics. While Mozambique has an abundance of water, the consistency of municipal supplies is reported to vary, even within the same municipality.

Chemical Waste Treatment

Just as the textile industries can require large amounts of clean water, depending on the extent and type of finishing, the effluent from these factories must be treated before disposal6. New investors in textiles depend on municipal facilities to treat waste water to the extent required. It is currently unknown if Mozambique's municipalities are capable of providing industrial waste water treatment. In the absence of proper waste treatment, untreated industrial waste is likely to end up in local rivers and water supplies, potentially damaging downstream industries such as fishing and tourism, not to mention a high potential to enter the local drinking water system and negatively impact human health in the short and long terms. Therefore, it is important for the Government of Mozambique to identify industrial sites within Mozambique that have access to proper waste treatment and to insure that standards are maintained. Where local waste treatment facilities are not capable of properly processing industrial waste, expensive upgrading of infrastructure will be needed.

Taxes and Incentives

As was reviewed earlier, textile firms possessing an export processing certification are eligible for a 60% reduction in corporate taxes for five years, reducing the corporate income tax rate from 32% to 16%. However, export processing certificates are only issued to companies with greater than 500 employees or ones that are located in established export processing zones with at least that many employees. Since textile firms are relatively capital intensive, modern spinning and weaving firms requires 100 to 300 workers, depending on the length of the working week.

As such, textile firms may not qualify for the standard tax benefits without modifications to the law. The Ministry of Industry and Trade (MIC) has proposed to the government that this provision be modified from 500 employees to 125 employees. This modification would be important to the development of an export led textile sector.

Mozambique's tax laws do permit the deduction of capital equipment costs up to an amount of 150% of actual cost, prorated over a period of five years.

Incentives given to global textile investors often include exemptions from income taxes for periods of 10 years or more. Greater incentives in the forms of infrastructure investments by local municipalities to support the manufacturing processes are also common (waste water treatment being common). Given the wide range of infrastructure between potential sites, no rule of thumb exists for incentive packages for textiles firms. Incentive packages are negotiated between governments and the investors. Subsidies in the form of electrical power and raw materials are also common, but no standards exist.

VERTICAL INTEGRATION AND END USE CONSTRAINTS

Processing cotton into yarns or fabrics without having to pay transportation and international logistics costs could reduce the costs of manufacturing textiles. Therefore, local access to cotton can be a significant benefit of locating a textile firm in Mozambique. But more than raw cotton is required. Availability of low cost electric, water, waste water treatment and access to transportation hubs and end users are all important and factor into an investors decision of where to locate a textile plant. So, while cotton is produced in the North of the country, industrial infrastructure and transportation are located in the South. Transportation between the North and South of the country is limited and shipping costs are as high as shipping from Mozambique to Asia or Europe. Investor will have to consider the benefits and costs of location carefully.

The existence of local downstream purchasers, such as apparel manufactures for fabric is a major advantage for a developing textile industry. With a population of less than 20 million people and a poverty rate of over 70 percent, the development of a local market for new textile and apparel products in Mozambique is limited, even with an import substitution strategy (compare the import substitution possibilities of China and India which have over one billion people each and poverty rates of between 10% and 25%).

With only one medium sized export apparel firm in the South of the country and limited potential for developing a local market, domestic purchasers for export quality yarns and fabrics within Mozambique are limited, as is the technical feedback needed for the development of high quality products.

COST OF FABRIC AND MATERIALS

APPAREL SUB-SECTOR CONSTRAINTS

Excluding the cost of fabric and materials, direct labour costs make up more than half of apparel firms non-material costs. Therefore, it is important that labour laws provide for flexibility in the production of garments. Mozambique's labour laws impose relatively high liabilities on producers in the form of paid leave, severance payments, and government's involvement in the hiring and firing processes. With Mozambique requiring severance and leave payments in excess of twice the regional and international average, it will be difficult to attract a large cluster of apparel manufactures, since labour costs are a major determinant of manufacturing costs and factor importantly in a producers decision of where to locate7.

The proposed labour law is significantly better for economic severance, at least in the short to near term. Regarding non-economic severance, after the third year of service, the proposed labour law is nearly identical to the old labour law. The new labour law is indeterminate about paid sick leave (resulting in high absentee rates). Finally, the government will still be highly involved in the dismissal of workers.

The rigidity of Mozambique's labour law posses the greatest constraint to the development of a cluster of apparel firms. The development of an apparel industry thereby limits the potential of upstream development of textiles and a major incentive to rehabilitate the cotton sector.

Shipping times

As reviewed earlier, apparel products have more in common with perishable produce then with metals or machinery. Reliable shipping times are of key importance to insure garments obtain their full value when sold. Mozambique, like all countries in Southern Africa, is a minimum of 4 weeks shipping time from major Asian textile producers. However, deep water cargo ships from Asia do not stop in any port in Mozambique. This requires cargo to be transshipped though South African ports introducing the possibilities of delays raging from 2 weeks or more as feeder ships await a critical backlog of containers to justify stopping in any Mozambique port.

Trucking may be used to reduce the variability of shipping times from Durban, but delays in border crossings from South Africa range from two to three days if all paper work is in order. Delays can be greater if there are discrepancies in paper work since border crossing communications are poor. Border crossing delays are largely viewed as resulting from the requirement that all cargo coming by truck pass through the Matola Cargo Terminal operated by FRIGO, a privately held company. Since FRIGO generates fees from container storage while clearing customs, a clear conflict of interest arises, since there is an incentive for FRIGO to question any shipment and incur delays, especially in the cases of high value shipments, such as fabrics and textile materials. Fees are based not only on the length of time containers await clearance, but also on the value of the cargo.

Also, in the case of customs clearance, many local actors express concern over the lack of competition in the despachante system of customs brokers, which operate virtual national monopolies on customs clearance processes. The lack of competition between despachantes results in lower levels of service and delays. Despachantes can take one week to clear cargo that would normally only take one to two days (Global Development Solutions 2005).

Taxes and Incentives

In today's apparel industries, competition among countries for foreign investors is extreme. As a result, it has become the international norm for apparel firms to be exempt from income taxes for the first five to ten years. Currently Mozambique offers a 60% cut on the normal 32 percent tax rate for the first 5 years of operations. After that, the normal tax rate applies.

With intense competition for foreign investors in the apparel industry, common incentives include subsidized buildings or factory shells, free trade zones with improved infrastructure and priority customs clearance within the zones. Mozambique currently does not offer subsidized or free factory shells. In the only free trade zone in the country, Beleluane, many building are reportedly awaiting construction due to high building costs. Other incentives for apparel firms may include skill development for operators or managers through vocational training programs, or matched funds for training.

MARKET ACCESS

In the past, countries in sub-Saharan Africa (SSA) could rely on consistent orders arriving on their shores as Asian producers became constrained by textile and apparel quotas in the major developed markets. Important changes are affecting global trade in textile and apparel products. These changes include:

- Elimination of textile and apparel quotas;
- Shifting preferential arrangements (including AGOA and other FTAs);
- Accession of China to the WTO in December 2001; and
- And the rise of trade remedies, such as safeguards and anti-dumping duties.

These changes are shifting the current imbalance in textile and apparel sourcing. Buyers in major markets used to source apparel products from producers with liberal quota access to developed country markets. Today, apparel buyers are sourcing from producers that are cost competitive, efficient and those that can guarantee the delivery of products. Although preferential programs can provide a needed incentive for the establishment of textile or apparel industries, Mozambique must still support the development of policies that are based on market fundamentals, such as efficient infrastructure, customs, regulations, and productivity rather than on preferential access. At the same time, Mozambique should consider the South African market and other regional markets as a destination for exports. Regional markets can provide much needed security and may hold greater potential than the traditional markets of the US and EU.

The scope of potential markets for Mozambique's textiles and apparel, therefore, should include not only the major developed markets (US and EU), but regional markets such as the Southern Africa Development Community (SADC),. Each of these markets, and their risks and potential.

Developments in the Major Markets (US and EU)

In the past, preferential trade benefits, such as the African Growth and Opportunity Act (AGOA) provided a set of incentives for bold and aggressive development of regional textile and apparel industries. The benefits of preferential trade include zero tariffs and near quota free access to the US and EU markets for qualified goods. The elimination of textile and apparel quotas on January 1, 2005 has had a negative impact on apparel exports from sub-Saharan African (SSA) countries including Mozambique. Moreover, the least developed country derogation under the AGOA agreement, that permits the use of non-regional fabrics and yarns will expire in September 2012, unless renewed by the US congress.

The window of opportunity for bold and aggressive steps to attract international producers has likely closed on SSA and Mozambique. The costs associated with attracting large transnational textile or apparel producers may not be justified based on market risks and fundamental costs associated with doing business in SSA, including long transit times to the major developed markets, labour regulations and excessive government red tape.

The advancement of US and EU regional trade agreements will also have a negative effect on Mozambique's access to major developed markets for apparel products. In the US, negotiations have been concluded for free trade agreements (FTAs) with five Central American countries and the Dominican Republic. Further negotiations promise to bring apparel suppliers from the Andean region, Thailand, South Africa (SACU) and North Africa.

The EU is also proceeding with its EuroMed program by inviting North African, Middle Eastern, Eastern European and Turkish producers into a PanEuro free trade area.

Free trade partners of the major developed markets seek to combine permanent duty free access to these markets with new rules of origin permitting the use of yarns and fabrics from within expanded regional blocks. The result will be a strong attraction for investment in these countries for the textile and apparel industries that might otherwise have located elsewhere. While the pace of these agreements and their impacts on SSA countries, such as Mozambique, are hard to predict, they do increase the number of duty free competitors to be confronted in those markets and will only increase competitive pressures and decrease the advantages of duty free access. They will also raise the bar on reducing lead times.

Protection measures, such as safeguards, anti-dumping (AD) and countervailing duties (CVD), will be on the rise and their effects will be uncertain and perhaps, short lived (the special textile safeguards applied on Chinese goods is set to expire at the end of 2008). While safeguards and AD/CVD actions may provide Mozambique an opportunity to export in the short-run, Mozambique should not depend solely on these measures to access markets as (1) such protection will end in 2008 and (2) capable and efficient suppliers are abundant and might not be part of AD/CVD actions.

Finally, if the AGOA third country fabric provision is not extended beyond September 2012, this would have a region-wide negative impact on exports and the demand for regional textile and apparel products. To be sure, a significant incentive to locate many garment firms in SSA is the ability to obtain duty free access on fabrics and yarns that are not regionally produced.

Therefore, a textile and apparel promotion program in Mozambique should take into account the regional availability of competitively priced fabrics and yarns that not only meet stringent international requirements, but are also in line with global prices.

Building links with competitive fabric and yarn suppliers in the SADC region could prove to be a significant competitive advantage to potential textile and apparel industries in Mozambique.

Trends in the world's textile and apparel markets dictate that Mozambique's producers will have to compete with other global suppliers on a level playing field.

African Growth and opportunity and Investment Incentive Act (AGOA)

The African Growth and Opportunity Act (AGOA) has provided sub-Saharan African (SSA) countries generous access to the US apparel market since 2000. Mozambique is included in this group of SSA beneficiary countries. A key provision of the AGOA legislation allows SSA countries (except South Africa and Mauritius) to processes imported fabric, from anywhere in the world. This so called third party fabric provision provides instant access to established global supply chains for the beneficiary countries. The provision has been considered a key benefit for many SSA countries to jump start their apparel industries, since the manufacturing of textiles in SSA is limited and is not expected to rise rapidly any time in the near future.

One of the last acts of the US Congress in 2006 was to pass the AGOA Investment Incentive Act of 2006. This act contains three provisions that are important to Mozambique:

- The extension of the third party fabric provision through September 2012;
- The inclusion of a rule that will limit the use of certain third country fabric that are in "abundant supply" in SSA;
- A provision extending duty free benefits to articles of textiles wholly formed in lesser developed SSA (excluding South Africa and Mauritius).

The extension of the third party fabric provision was met with mixed enthusiasm by textile and apparel producers. On one hand, most SSA apparel producers depend on imported fabrics for the apparel they export to the US. On the other hand, that access to third party fabrics discourages investment in the SSA textile industry.

The development of a textile industry still lags behind apparel exports in its ability to supply fabric. The second major provision in the AGOA Investment Incentive Act addresses the concern over the development of a regional textile industry, by closing the door to third party fabrics when local materials are available or are judged by the US International Trade Commission (USITC) as being in commercial supply.

Finally, the third provision of the AGOA Investment Incentive Act extends duty free benefits to articles of textiles, such as home textiles, yarns and fabrics if the products are wholly formed in lesser developed SSA countries. A wholly formed product must be made from SSA fiber on up to the final product, a much stricter standard than that for apparel, which does not require the fiber to be from SSA.

Mozambique could consider attracting a vertically integrated (cotton to product) home textile producer that could meet this requirement establishing a new market niche under AGOA.

With the recent extension of the AGOA third party fabric provision for six more years, the greatest constraint on Mozambique will be its own ability to make the internal reforms needed to attract apparel investors. The window of opportunity will again close, but probably sooner than 2012, when the third party fabric provision expires.

Preferential Access to the EU Market

While the passage of new AGOA legislation put aside uncertainty over Mozambique's preferential access to the US market; preferential access to the EU market is still uncertain. Today Mozambique can access the EU market via two preferential trade agreements:

- The African, Caribbean and Pacific Islands (ACP)/Cotonou Agreement; and
- The EU GSP program for least developed countries—Everything But Arms (EBA).

The EU ACP/Cotonou agreement is more generous than the EU EBA program in that it provides for the use of regional textiles for apparel benefiting from duty free treatment exported to the EU (although South African textiles have never been approved for eligibility in apparel benefiting from duty free access). In contrast the EU EBA does not permit the use of regional or third country fabrics requiring apparel exported from Mozambique to include fabrics knit or woven in Mozambique.

Of considerable importance, the ACP/Cotonou Agreement is set to expire in 2008. As a direct consequence, ACP countries are expected to negotiate and conclude Economic Partnership Agreements or EPAs with the EU before the expiration of the ACP/Cotonou agreement. These EPAs are not unilateral like the ACP agreement, and are full free trade agreements (FTAs) requiring reciprocal market access provision. The process of negotiating these EPAs is far behind schedule and the EU's intentions are just being established. Two points that are being negotiated today give an early indication of new provisions to come:

- ACP countries are being organized into regional blocks, where supply chains will be linked (cumulation provisions); and
- New value added rules of origin are being explored as an alternative to the current EU systems of lists and transformation criteria.

The EU has designated six regional trade blocs for negotiations and Mozambique is include in the "Southern Africa—SADC" trade block. The fact that this block excludes South Africa and includes SACU countries raises numerous questions of how this trade block will work. South Africa already has a free trade agreement with the EU, but is only an observer to the SADC-EPA negotiations. SACU countries are actively engaged in the EPA negotiations, but as members of SACU, are more closely aligned with South

Africa. SACU countries may seek their own FTA with the EU. The EU has already signaled its desire that SADC strengthen its role as a regional trade block to implement the EU-EPA. So, it may be that the biggest impact on Mozambique of the EU-EPA will be the strengthening of the SADC community. This would have important implications for any Mozambique textile or apparel industries, since it would mean tighter regional integration with SADC supply chains from cotton, yarn, fabric to apparel.

Mozambique would do well to actively participate in this process to maximize its access to the EU market, through the use of regional textiles and fabrics on an unrestricted basis.

Today, the use of regional textiles in apparel benefiting from tariff free access to the EU is limited, since South African textiles are notable excluded from materials permitted preferential access. Moreover, Mozambique's access to SADC markets is notably reduced because SADC rules of origin require two stages of transformation (yarn spinning and fabric forming, or fabric forming and garment making) for any material or product to be eligible for duty free treatment within SADC—a condition which greatly limits Mozambique's advantage as a low cost labour country that is potentially competitive in apparel with limited opportunities in the capital intensive textile industry. Therefore, Mozambique has much to gain from the reorganization of SADC and the alignment of the trade block with the EU.

Southern Africa Development Community (SADC)

With long shipping times and limited textile capacities, SADC's largest member state, South Africa, will offer perhaps the greatest potential to Mozambique's development of an apparel industry and eventually a textile base—be it as a source of materials or investors looking to produce these goods more competitively in a different location.

However, today, Mozambique's preferential access to the South African market, via SADC, is limited. The constraint being the SADC rule of origin for textiles and apparel that requires two stages of transformation for a good to be conferred originating status.

In this case, apparel cut and sewn in Mozambique of fabric formed in South Africa is not eligible for duty free status unless the South African fabric is also made of yarn spun in South Africa. The key ingredient that is lacking is a provision for "cumulating" of origin across SADC countries. In the case of cumulating, SADC would require two stage of transformation in any SADC country be counted towards originating status. In this way, fabric formed in South Africa, and sewn in Mozambique would be eligible for duty free treatment anywhere within SADC.

It is important to recognize the role of the EU-EPA negotiations and their potential to transform the SADC region and the rules of origin that constrain

SADC trade. Mozambique would do well to monitor the EU-EPA negotiations for any opportunity to not only improve their access to the EU market, but to further the integration of SADC textile and apparel industries and markets.

Finally, under SADC, Mozambique is granted a small quantity of apparel exports to South Africa under the MMTZ quota for apparel that is constructed of third part fabrics.

Bibliography

A. Brand, J.P.T.M. Noordhuizen and Y.H. Schukken: *Herd Health and Production Management in Dairy Practice*, IBDCO, 2003.

A. Sahoo and R.K. Singh: *Animal Health and Production*, Satish Serial Publishing House, 2013.

A.J. Solomon: *Advances in Pollen-Spore Research: Vol. 29. Latest Approaches in Pollen Biology and Reproduction*, Today & Tomorrow's Printers and Publishers, 2011.

A.P. Gupta, G.P. Rao and B.L. Srivastava: *Cane Sugar Production Management*, IBDCO, 2000.

A.R. Prasad: *A Study on the Theory of Production*, Reliance Pub, 2005.

Abhinash Kumar: *Material Management*, Rajat Publications, 2011.

Ajoy Gangopadhyay: *Assessment of Improved Crop Production Technologies*,Gene Tech Books, 2007.

Anubhav Dwivedi: *A Fragrance from Food Production World*, Axis, 2010.

Ashok Kumar Sharma: *Animal Reproduction*, Oxford Book Company, 2012.

Bhagaban Swain: *Adoption of New Technology : Production Efficiency and Agrarian Relations*, Kalpaz, 2002.

C. Gangaiah: *Area Production and Productivity of Horticulture Crops*, Serials Publications, 2011.

C. Venkatesan and D. Namasivayam: *Agricultural Production in India*, Global Research Pub, 2011.

Chandra Shekhar: *Arming the Defence Forces : Procurement and Production Policies*, Manas, 2004.

D N Singh and S P Singh: *Production Management of Underutilized Vegetables*, Agrihortica Pub, 2007.

Dave Ingram:*Applied Practices And Principles For Production-Ready Software Development:Design-Build-Run*, Wiley.

David Ludden: *Agricultural Production and South Asian History*, Oxford University,

2005.

I.S. Singh: *Aonla : Production Handling and Processing*, Westville Publishing House, 2012.

Jaya Banerjee: *Production Management : Modern Methods and Techniques*, Shree Niwas Pub, 2011.

K. Vanangamudi, N. Natarajan, P. Srimathi, K. Natarajan: *Advances* in Seed Science and Technology, Vol. II, Quality Seed Production *in Vegetables* , Agrobios, 2006.

K. Vanangamudi: *Advances in Seed Science and Technology : Vol. VI: Fruit Seed Production*, M. Prabhu and A. Bharathi, Agrobios, 2011.

M.A. Shewan: *Production Management*, Sonali, 2008.

N.N Sarkar: *Art and Print Production*, Oxford University Press, 2008.

Nirmal Sengupta: *A New Institutional Theory of Production : An Application*, Sage, 2001.

Okoro M. Akinyemi: *Agricultural Production : Organic and Conventional Systems*, CRC Press, 2013.

P. C. Sharma: *A Textbook of Production Technology : Manufacturing Process*, S. Chand Publisher, 2008.

P.K. Dhawan: *Research Methodology in Material Management*, ALP Books, 2011.

P.L. Saroj and O.P. Awasthi: *Advances in Arid Horticulture, Vol. II : Production Technology of Arid and Semiarid Fruits*, International Book, 2006.

Parvinder S. Bali: *International Cuisine and Food Production Management*, Oxford University Press, 2012.

Professor Jack Milton Widholm Commemorative Volume : *Applications* of Plant Biotechnology : In Vitro Propagation, Plant

Transformation and Secondary Metabolite Production , I.K. International Pub, 2010.

R.K. Chauhan: *Production Management*, Book Enclave, 2005

S P Singh: *Production Management of Spices*, Agrihortica Pub, 2007.

S. Kannaiyan and K. Kumar: *Azolla Biofertilizer for Sustainable Rice Production*, Daya, 2005.

S. Prasad and U. Kumar: *A Handbook of Fruit Production*, Agrobios, 2010.

S.K. Verma: *Material Management*, ABD Pub, 2011.

S.P. Mittal: *Alternative Land Use Systems for Sustainable Production*, Oriental Enterprises, 2004.

Sampat Nehra: *Arbuscular Mycorrhizae in Crop Production : Professor P.C. Trivedi Festschrift Volume*, Pointer Publishers, 2013.

Shamsher S. Narwal: *Allelopathy in Crop Production*, Scientific, 2004.

Index